# Non-native Marine Species
## in the
# Channel Islands

*- A Review and Assessment -*

**Department of the Environment**

- 2017 -

States
of Jersey

Non-native Marine Species in the
Channel Islands: A Review and Assessment

First published in Great Britain
in 2017 by Société Jersiaise
societe-jersiaise.org

ISBN 978 0 901897 13 8

*Hvis man fortier et spøgelse, vokser det sig større.*

# Contents

# 6 - Summary and Recommendations

# 7 - Recomendations for the Channel Islands    207

# 8 - Endogenic, Cryptogenic and Exogenic Species

# Preface

Non-native species are plants, animals and other organisms that have been found living outside of their natural geographic range. Such species are capable of having a serious negative impact on the environment, economy and even human health. The Convention on Biological Diversity rates non-native species as being the second greatest threat to the global environment after habitat loss. All parts of the world are affected by non-native species as are most ecosystems within the terrestrial, freshwater and marine realms.

The Channel Islands are British Crown Dependencies located off the Normandy coast in the western English Channel. The islands are at a geographic crossroads between colder marine waters to the north and warmer waters to the south. This makes them suspectable to an unusually wide range of non-native marine species which may reach them from points of origin to the north and south.

The past decade has seen a marked increase in the number and abundance of non-native marine species being recorded from the Channel Islands. Awareness of the potential threat posed by these organisms has been growing regionally but information networks and research projects concerning marine non-native species has lagged behind that of the terrestrial and freshwater world.

The report offers a review and assessment of the non-native marine species situation within the Channel Islands. The report was produced by the Marine Resources Section within the Department of the Environment (States of Jersey) and is part of a wider non-native species strategy that will cover the whole island. Although this report focuses principally on Jersey and its offshore dependencies, an attempt has been made to assimilate and synthesise available information from the other Channel Islands as well.

The objectives of this report are to: assess the current and historic occurrence of non-native marine species in the Channel Islands; horizon scan for species reported from neighbouring regions; analyse these data to assess the behaviour, rate of spread and threats posed by these species; and to make basic observations and recommendations regarding the current and future status of non-native species in the Channel Islands.

The information from this study will be integrated into the wider non-native species strategy that is being undertaken by the States of Jersey. For this reason, only basic conclusions have been included here as the results will need to be subjected to additional analysis and scrutiny as part of the strategic reporting process.

It is hoped that this report will act as a useful introduction to non-native marine species in the Channel Islands area and to the issues associated with them. It may also provide a baseline for future studies of non-native marine species in Jersey and the other Channel Islands.

# - Part One -

## Introduction and Background

# 1 - Background

Non-native species are organisms that are living outside their natural distributional range. They are known by many different names including: alien; exotic; introduced; invasive; and non-indigenous. This report uses the term 'non-native' which refers to species that have been accidentally or deliberately introduced into an area via human activity.

Species that have a measurable negative impact on a local environment, health or economy are someitme referred to as being 'invasive non-native species'. However, this label (while valid) is sometimes confused with more general references to 'invasive species' and so to avoid such confusion this report will avoid using the term invasive where possible.

## 1.1 - Non-native Species: A Definition

Wxactly which criteria should be used to define what is (or is not) a non-native marine species varies between researchers. Some include any intentionally or unintentionally introduced species regardless of its locus of origin, time of arrival or ability to reproduce locally while others are more prescriptive.

This report is primarily interested in those non-native marine species that have been deliberately or accidentally introduced into north-west Europe from elsewhere and which have the potential to establish a breeding population in Channel Island waters.

This largely excludes those species that might find their way to the islands but are not able to reproduce usually because of unsuitable environmental conditions such as sea temperature (e.g. turtles, tropical barnacles, shipworms, etc.). This also excludes most cryptogenic species, i.e. those plants and animals whose origin is unknown or uncertain. These are often species that have not been intentionally or acccidentally introduced by humans but whose natural distribution range has extended allowing them to enter new areas. A list of possible crptogenic species which fall into this actegory is given in Chapter 8.

The species that are of most concern to this report fit into the schemes given by Minchin *et al.* (2013) and Wolff (2005) which hold that to be a non-native organism a species must fulfil one or more of the following criteria:

- A taxonomically distinctive species which has no previous recorded history in a region.
- A distinct geographic gap between the European occurrence of a species and the remainder of its population.
- A species with a highly localised occurrence.
- A localised species whose initially restricted range suddenly expands.
- A lack of any obvious natural pathway/vector between individual species

populations.
- A species whose population is rapidly expanding.
- A species which is associated with a known artificial means of translocating such as aquaculture, shipping, etc.
- A species that is parasitically dependent on another non-native species.
- Low genetic variability within a species population.
- A genetic similarity to a geographically distant species population.
- A species with a life history stage that cannot easily be dispersed by natural means.

## 1.2 - Methods of Introduction

A common characteristic of non-native marine species is the ability to spread rapidly or to be reported from two or more isolated locations but without being recorded from anywhere in-between. For a non-native species to enter a new area and then spread successfully requires the establishment of a pathway between locations and a means of transport. These are often referred to as being the 'vector' by which a species has moved from one location to another and, in the case of non-native organisms, these vectors are usually associated with human activity.

For non-native terrestrial species there are many types of vector (from the soles of travellers' shoes to freight containers) but in the marine realm these are fewer in number. This section will list those marine vectors that are most relevant to non-native species found on the Channel Islands.

### Shipping

Non-native marine organisms may be transported by ships and boats usually attached to their hulls or, if planktonic or at a larval stage, in ballast tanks. When a ship docks in a harbour or flushes its ballast tanks at sea, non-native species may enter the local environment and, if conditions are suitable, begin to disperse and reproduce. Such organisms will often initially establish themselves in a port or harbour and disperse from there either into the local marine environment or on other boats travelling to neighbouring ports.

The Channel Islands are situated close to the central English Channel which has one of the world's busiest shipping lanes. This area accommodates cargo and other boats arriving from ports and harbours from across the entire globe. This places English Channel ports at a high risk of receiving non-native species translocated from foreign sea areas via shipping. Once established at one English Channel port, a non-native organism may soon be recorded at neighbouring ports and harbours but without being found on the coastline in-between. This disjointed pattern of distribution is often a sign that shipping is the main transport vector for a species. (For a detailed analysis of UK shipping vectors see Tidbury *et al.*, 2014.)

The Channel Islands have continual commercial and leisure shipping links to each other and to medium and large-sized ports on the coast of southern England, Normandy and Brittany. The islands also have regular

*Figure 1.1 - Marine transport is essential to the economy and well-being of the Channel Islands but it can also assist with the spread of non-native species.*

links to other ports and harbours in Europe via cargo boats, yachts, cruise ships, naval vessels, etc. It is therefore feasible for a for non-native marine species to move between the Channel Islands and local and regional ports which probably explains why there is an overlap between the non-native species found in St Helier and St Peter Port and those recorded in ports at Brittany and southern England (see Figure 1.3).

## Aquaculture

For over a century the coastlines of Normandy and Brittany have had large onshore and offshore aquaculture industries. This is especially true in southern Brittany and Biscay whose aquaculture areas have been a historical point of entry into Europe for several non-native species.

The principle aquaculture vector for non-native species is the movement of seed stock which, if imported from abroad or relocated within a region, may carry other organisms with it. A second vector may be the farmed species themselves, some of which will reproduce and become established in areas neighbouring aquaculture concessions.

Within the Channel Islands there are aquaculture industries on Jersey and Guernsey but the importation of seed stock into both islands is subject to strict biosecurity measures which lessens the threat from associated non-native species. However, two of the farmed species (Manilla Clam and Pacific Oyster) are found on the seashore and may have been introduced via local aquaculture.

The aquaculture industry within the wider Normano-Breton Gulf and adjacent areas dwarfs that of the Channel Islands and historical absences/ lapses in biosecurity (usually associated with importation of seed stock) have led to arrival of many non-native species, especially in the Bay of Biscay. Historical aquaculture activities have been a major source of non-native species to the western English Channel region and therefore also a

*Figure 1.2 - This could be the moment when the Manilla Clam (*Ruditapes philippinarum*) was introduced into Jersey through a States of Jersey aquaculture trial in St Catherine's Bay in 1986.*

major (if indirect) vector for species entering the Channel Islands.

## Secondary Dispersal

Once established at a location (such as a port or aquaculture area) some non-native species will begin to spread outwards into the neighbouring marine environment. These species may have no predators or other natural checks to their growth and reproduction allowing them to out-compete native organisms. In doing so they may alter habitats, biomes and trophic webs, all of which can have knock-on effects for the natural marine environment and even for ecosystem functions (such as the economy, health, etc.) that are associated with human society.

Secondary dispersal often occurs during the planktonic phase of an organism's lifecycle although not all species will have this. Eggs and early stage larvae may be carried by tidal currents for long distances before settling onto the sea floor to grow into adults. For offshore reefs and islands, this can be a major vector for non-native organisms that are established on neighbouring coastlines.

Computer modelling of long-term residual tidal currents within the English Channel suggests that seawater moves west to east from the Atlantic Ocean towards the Straits of Dover (Figure 2.2). The Channel Islands, being located near to entrance of the English Channel, are dominated by tidal currents originating from its western approach. Conversely, the islands are

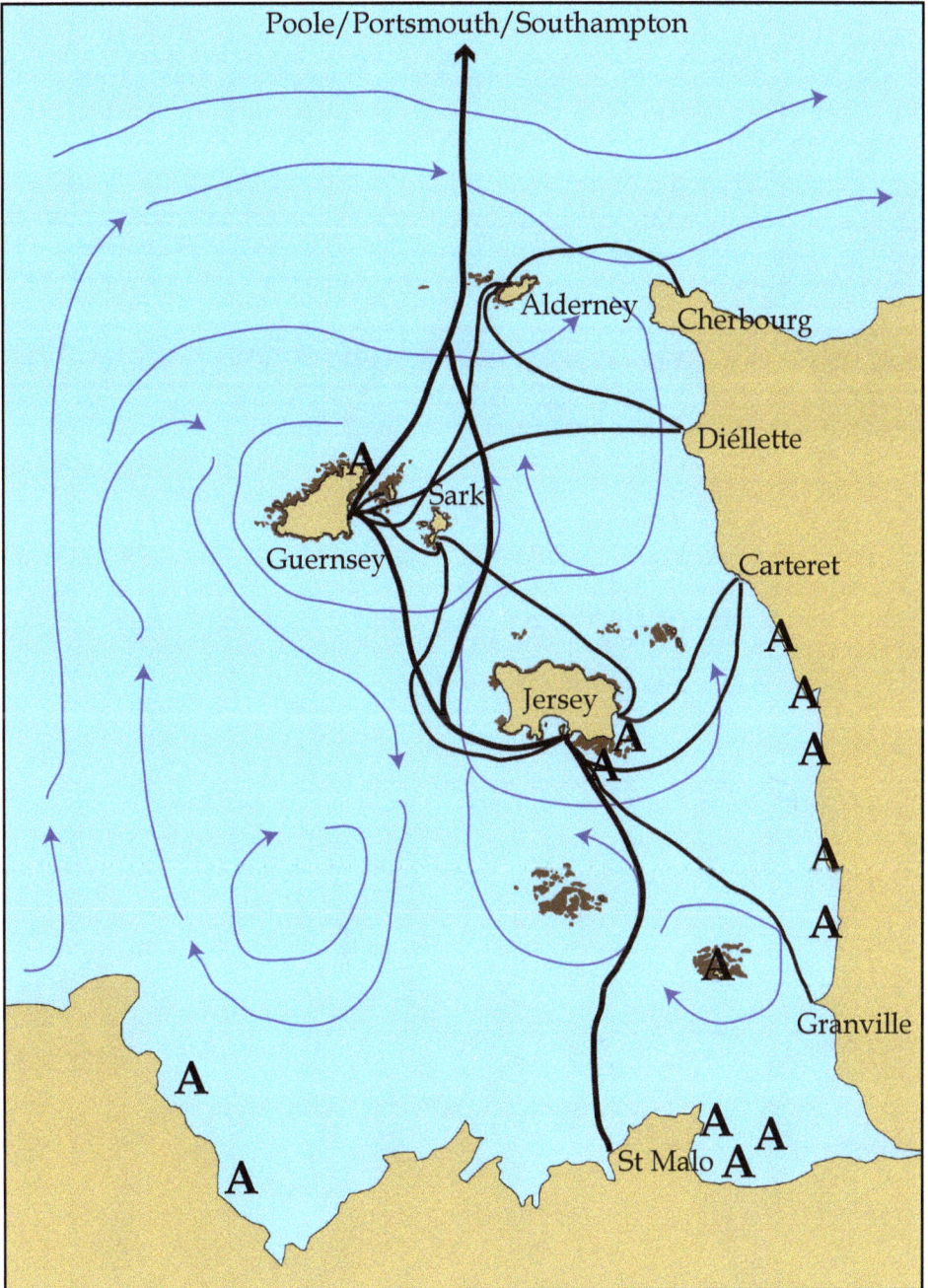

*Figure 1.3 - Potential transport vectors for marine non-native species between the Channel Islands and neighbouring coastlines.*
**Black lines** = *The principal commercial ferry and cargo shipping routes entering and leaving Channel Island ports and harbours.*
**Blue lines** = *Residual tidal circulation within the Normano-Breton Gulf (After Jegou and Salomon, 1990).*
**A** = *Important shellfish aquaculture areas.*

less influenced by tidal currents from the eastern Channel. It is therefore more difficult (but not impossible) for organisms living in the eastern English Channel area to enter the Normano-Breton Gulf via tidal currents. To do so they usually spread westwards along the north Normandy coast and around the Cherbourg Peninsula.

The central English Channel also presents an impediment for any species (including non-native ones) crossing from the southern English coast direct to the Channel Islands. Tidal currents, water depth and a lack of sedimentary habitats creates a natural barrier that separates the Normano-Breton Gulf from the United Kingdom. However, it is possible for non-native species to cross the Channel further east, where it is narrower and shallower, and then spread west along the French coast towards the Normano-Breton Gulf. Based on current observations, it appears that non-native species spreading west from the Straits of Dover (such as the Jack-knife Clam) cannot readily round the Cherbourg Peninsula. This may offer the Channel Islands some protection or a delay against some species becoming established locally.

Within the Normano-Breton Gulf a complex arrangement of local gyres and residual currents aid the transport of planktonic larvae to the Channel Islands from the north Brittany and lower Normandy coasts. Once a species enters the Gulf, it can spread rapidly along mainland coasts and then offshore to individual reefs and islands.

Outside of regional tidal currents (which tend to affect larvae and planktonic organisms), adult animals and plants may move along a

*Figure 1.4 - The Normano-Breton Gulf has some of the strongest tidal currents in the world which, during spring tides, may reach several knots. These currents aid the dispersal of planktonic species and larvae within the Gulf including to offshore areas such as the Channel Islands.*

coastline under their own volition by walking or swimming. Physical dispersal of this sort spreads outwards from a point of origin and may happen in any direction and at variable speeds, depending on the organism concerned. The direction and rate of spread tends to be controlled by the biological and environmental requirements of the species concerned and the availability of suitable habitats into which it can spread, rather than water circulation.

When looking at the regional situation, the Channel Islands are vulnerable to non-native species naturally dispersing along any part of the English Channel although natural barriers to the north and east make it more likely that they will spread from other parts of the Normano-Breton Gulf than from the eastern English Channel or southern English coastline.

## Other Vectors

Outside of shipping, aquaculture and secondary dispersal, transport vectors of lesser or no importance to the Channel Islands will include dry ballast, which was not widely used after the 1870s, aquarium escapees and introductions via freshwater systems such as canals.

## 1.4 - Threats Posed by Non-Native Species

Many non-native species arrive into environments where they have no natural predators or diseases. Coupled with this rapid growth and an effective reproduction strategy, a non-native species can have a devastating effect on the native habitats and species that it encounters.

Non-native species have a reputation for breeding and spreading at such a rapid rate that they dominate individual habitats, displacing native species and changing the ecological balance around them. Examples of this include Wireweed (*Sargassum muticum*) and the American Slipper Limpet (*Crepidula fornicata*) both of which have impoverished key habitats within the Channel Islands since the 1970s. Fortunately only a few non-native marine species have so far had such an extreme effect but most will affect native organisms and habitats to a greater or lesser degree. For this reason non-native species (terrestrial, freshwater and marine) are usually listed as a major cause of global biodiversity loss along with more well-known issues such as over-development, deforestation and pollution.

Sometimes the effect of a non-native species is more limited. This is particularly true of pathogens and parasites which are usually host specific but can nonetheless have dramatic consequences. For example, the spread of the pathogen *Bonamia ostreae* had a devastating effect on the cultivation of oysters across Europe in the 1980s, almost wiping them out in some areas.

Other species, particularly phytoplankton, can form toxic blooms that will kill fish, shellfish and even larger animals. This phenomenon is usually associated with estuarine conditions and is unlikely ever to become an issue for the Channel Islands but it has affected European shellfish farms.

The ability of some encrusting non-native species to dominate substrates can lead them to becoming a fouling nuisance, especially on boat hulls, pontoons, pilings and even water intakes. This is particularly true of bryozoans, barnacles, ascidians and bivalve molluscs, with the principal consequence being the cost of keeping surfaces clean of organisms. Fouling most often occurs in harbours where any removal costs are usually borne by ports authorities and boat owners. However, some species, such as the Pacific Oyster (*Crassostrea gigas*), Asian Bryozoan (*Watersipora subatra*) and Carpet Sea Squirt (*Didemnum vexillum*), can foul small areas of open coastline too leading to a loss of biodiversity.

Other threats include the disruption of beneficial ecosystem functions such as nutrient recycling and the effectiveness of food chains. Some species may also have an effect on local genetics through hydridisation (e.g. the Canadian Lobster) or a reduction in the local gene pool through loss of biodiversity and abundance. A full assessment of the potential threats posed by non-native marine species in the Channel Islands area is given in Chapter 2.4.

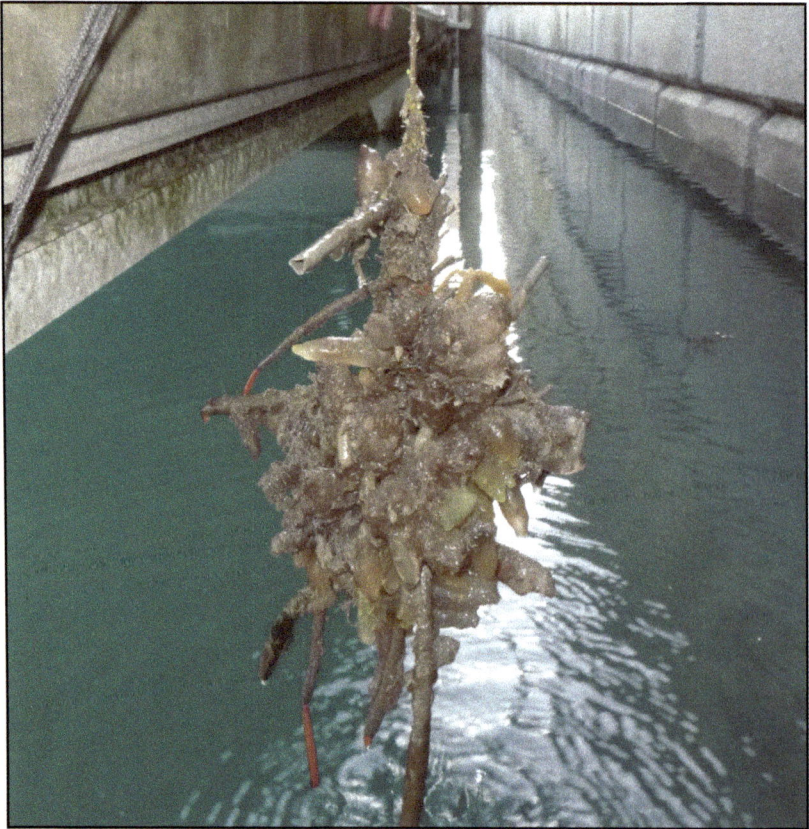

*Figure 1.5 - A rope left dangling in St Helier Marina, Jersey, has been colonised by sea squirts, bryozoans, sponges, tube worms and other fouling organisms.*

## 1.5 - Management and Legislation

For several decades there have been attempts at controlling and managing the potential biological, economic and sanitary effects posed by non-native marine species. Measures have included legislation, restrictions on aquaculture stock movement and physically removing established plants/animals from the environment. None of these techniques has been completely successful and in some circumstances may even have made the situation worse.

When looking at management options for non-native marine species, there is only one viable option: preventing the species from arriving in the first place. This has been recognised for many years but for prevention to be effective it requires coordinated action between every country with access to a coastline and the cooperation/coercion of all international shipping, aquaculture and other industries. So far attempts at cross-border prevention have only a partial track record of success.

At an international level there is legislation concerning the prevention, reduction and management of non-native species in the 1982 UN Convention on the Law of the Sea. In September 2017 the International Convention for the Control and Management of Ships' Ballast Water and Sediment came into force. Both these conventions are primarily designed to prevent (or at least minimise) the transport of non-native species via commercial shipping.

There are also a number of international environmental conventions which recognise the threat posed by non-native marine species and which prioritise their prevention and reduction. This includes the Convention on Biological Diversity (CBD), and also the Ramsar, Bonn, Bern and OSPAR conventions. Jersey is an annexed signatory to the CBD, Bonn, Bern and OSPAR conventions and Ramsar sites exist on the larger Channel Islands.

At a European level there are a number of EU directives on the environment, aquaculture and water quality that cover non-native species. Probably the two most important ones are the Marine Strategy Framework Directive (2008/56/EC) and Regulation 1143/2014. Both these deal directly with the issue of non-native species and have come into force.

Although preventing the arrival of non-native species is the most effective management mechanism, it requires a level of cooperation and adherence to regulations which is difficult to achieve. Even effective management systems cannot cater for every eventuality and so there will always be new non-native species being introduced to western Europe.

Once a non-native species has established itself in the marine environment, the next priority is to prevent its secondary dispersal to other locations. Eradication is an obvious option and, while this has worked with some terrestrial species, it is often logistically difficult and expensive to accomplish. In a wide, pervasive environment such as the sea, eradication is almost impossible unless achieved soon after the species' arrival. There are no examples of a marine non-native species having been successfully eradicated in Europe although some have died out naturally or been controlled locally.

The confinement, control and reduction of individual species is also an option. Here there has been some success but generally only with species-specific parasites that present an economic threat to aquaculture or human health. Many countries have introduced biosecurity measures which restrict or prohibit the movement of aquaculture stock or demand that seed stock is sourced from national rather than international locations. Similar restrictions have been enacted in the Channel Islands and, while primarily aimed at preventing the entry of pathogens such as *Bonamia ostreae*, they also help to prevent the entry and secondary dispersal of other non-native species with seed stock.

When it comes to the management of non-native marine species, small regional areas such the Channel Islands are primarily fighting a battle against secondary dispersal from neighbouring coasts. The biosecurity measures associated with aquaculture in Jersey and Guernsey have been successful but beyond this there is little that the islands can do to prevent larvae drifting into their waters or to stop vessels arriving from France, the UK and elsewhere.

A viable management option for the Channel Islands is to make sure that they are actively monitoring and evaluating the risk from established non-native species and are aware of those species that might arrive in the near future. Further research, action plans and possibly some physical management may be needed for individual species but in almost all instances, rolling back the arrival and spread of an established species will be impossible.

*Figure 1.6 - Oyster farming in Grouville Bay, Jersey. Strict biosecurity measures are essential for the sake of the aquaculture industry and the local marine environment.*

# - Part Two -

## A Survey and Threat Assessment

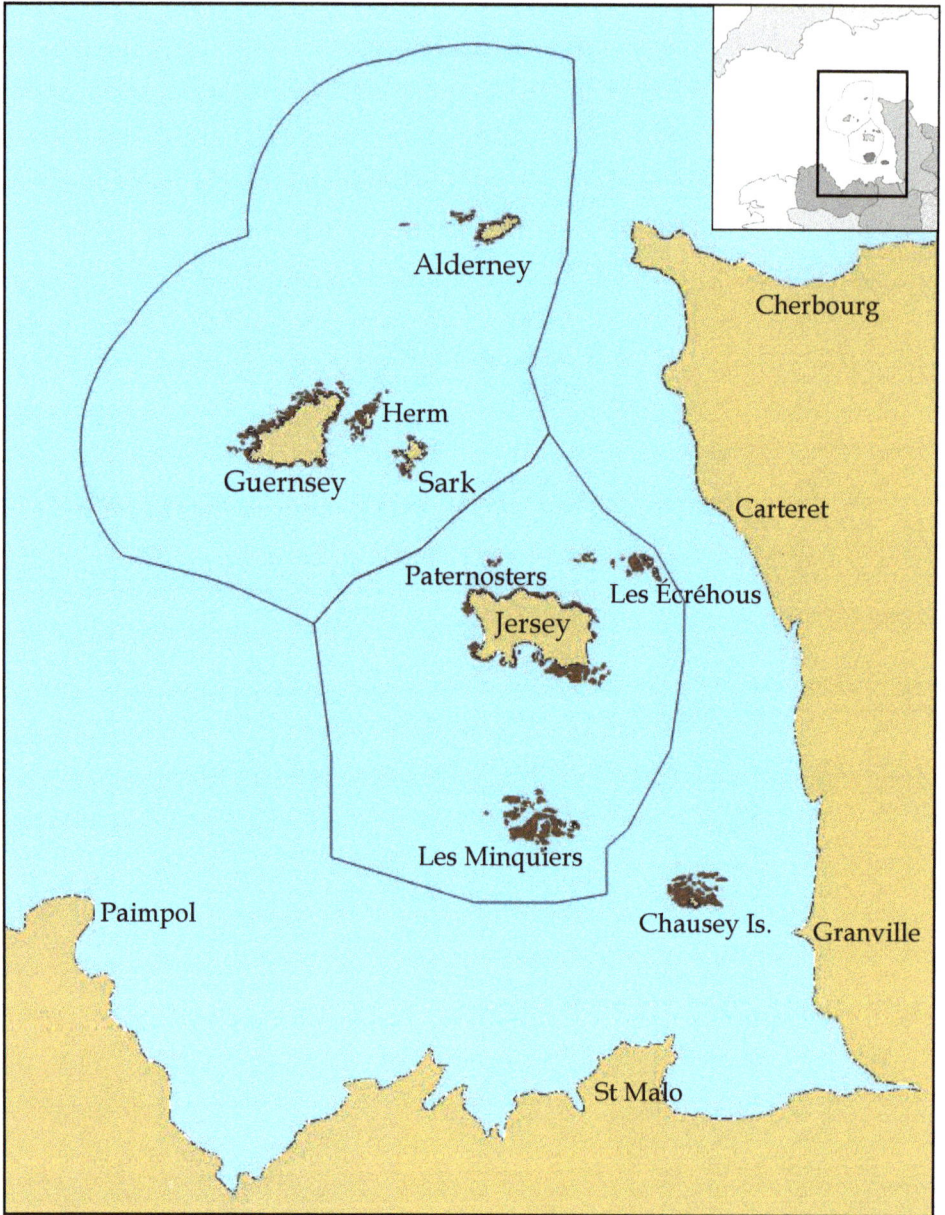

*Figure 2.1 - The Normano-Breton Gulf and the Channel Islands. The territorial seas of the Bailiwicks of Jersey and Guernsey are marked as are the principal islands, reefs and French coastal towns.*

# 2 – Survey Area and Methodology

During the past decade comprehensive lists of non-native marine species have been compiled for several European countries. The creation of these lists has been essential when quantifying the potential threat presented by non-native species to a local area or region.

Within the Channel Islands some basic lists of non-native marine species have been compiled but these are localised to individual islands and only utilise local biological records (see Chapter 2.2. below). Understanding the collective threat posed by non-native marine species requires baseline knowledge from a much wider area.

In 2016 a full census of non-native marine species was undertaken using original research in combination with a survey of literature, databases, archives and other published and unpublished sources. This survey relates not just to marine non-native species in the Channel Islands but also the English Channel, Atlantic European coastline and southern North Sea. The primary objectives of this survey were:

1 - To compile a list of non-native species known to be (or have been) present within the Channel Islands territorial sea area.

2 - To compile a list of non-native species reported from sea areas neighbouring the Channel Islands.

3 – To assess potential threats posed by these non-native species to the Channel Islands and, if not present already, the likelihood of a non-native species reaching and establishing itself in the islands.

This chapter defines the survey's parameters and methodology including a discussion about information sources. The results and analyses are presented in Chapters 3 and 4.

## 2.1 – The Survey Area

The Channel Islands are located in the Normano-Breton Gulf at the western end of the English Channel between the French coasts of Normandy and Brittany (Figure 2.1). This sea area is noted for its large tidal range which can reach 9.8 metres in Guernsey, 12.2 metres in Jersey and 13 metres in the Bay of Mont St Michel.

The Channel Islands are British Crown Dependencies which means that they are a part of the British Isles but not part of the United Kingdom. The islands are formed of two self-governing territories:

1 - The Bailiwick of Jersey which includes the island of Jersey and three uninhabited offshore reefs called Les Écréhous, Les Minquiers and Les Pierres de Lecq (Paternosters).

2 - The Bailiwick of Guernsey which includes the populated islands of Guernsey, Alderney, Sark and Herm plus several smaller islands and islets that are sparsely/seasonally populated or uninhabited.

The islands have their own territorial waters which jointly cover 6,223 km² (Jersey = 2,455 km²; Guernsey = 3,768 km²). Culturally and politically the islands reflect aspects of their Norman and English heritage although the latter has become more dominant since World War II.

The shape and location of the Normano-Breton Gulf generates strong tidal currents which flow around the Channel Islands in a series of gyres, creating a residual circulation pattern that may retain seawater locally for days or weeks (Fig. 2.2). This, in combination with the natural barrier formed by the central English Channel (see Chapter 1.2), prevents some marine species from dispersing from the Gulf to the north and east (see Hir *et al.*, 1986; Greenaway, 2001; Chambers *et al.*, 2016).

An inability to disperse further into the English Channel makes the Normano-Breton Gulf the northernmost limit for a number of southern European species, such as the Ormer (*Haliotis tuberculata*), Pennant's Topshell (*Gibbula pennanti*) and Mint Sauce Worm (*Symsagittifera roscoffensis*). It is estimated by the Société Jersiaise that around five per cent of marine species recorded from Jersey are not found on the United Kingdom coast. The same natural barrier also prevents some marine plants and animals moving south across the English Channel from the coast of the United Kingdom to the Normano-Breton Gulf (see Chapter 1.2; Holm, 1966).

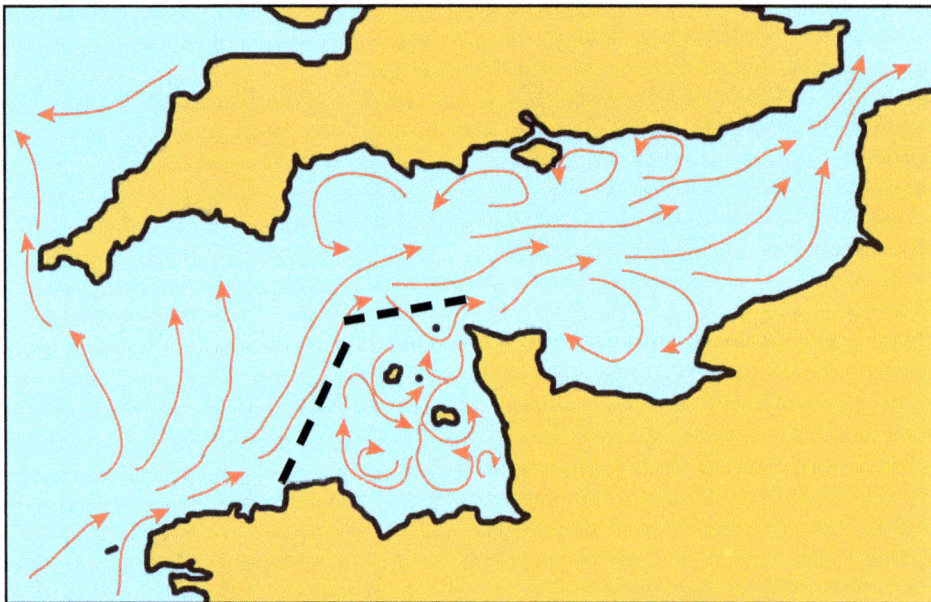

*Figure 2.2 - Long term water circulation within the English Channel and Normano-Breton Gulf areas. The dashed line illustrates a natural barrier which discourages some species from dispersing away from the Normano-Breton Gulf and into the central and eastern English Channel. (After Jegou and Salomon, 1990)*

However, the Normano-Breton Gulf is not entirely cut off from the English Channel as species may enter from the east via the Cherbourg Peninsula having travelled along the coasts of Holland, Belgium and France. Marine species may also enter the Normano-Breton Gulf from the Bay of Biscay via the Finistère peninsula.

With connections to the colder Boreal waters of northern Europe and the warmer Lusitanian seas of Biscay and the Iberian Peninsula, the Normano-Breton Gulf is an area of notably high biodiversity and productivity. This means that the Gulf can potentially receive non-native species from an unusually wide geographical area. The connection of the Normano-Breton

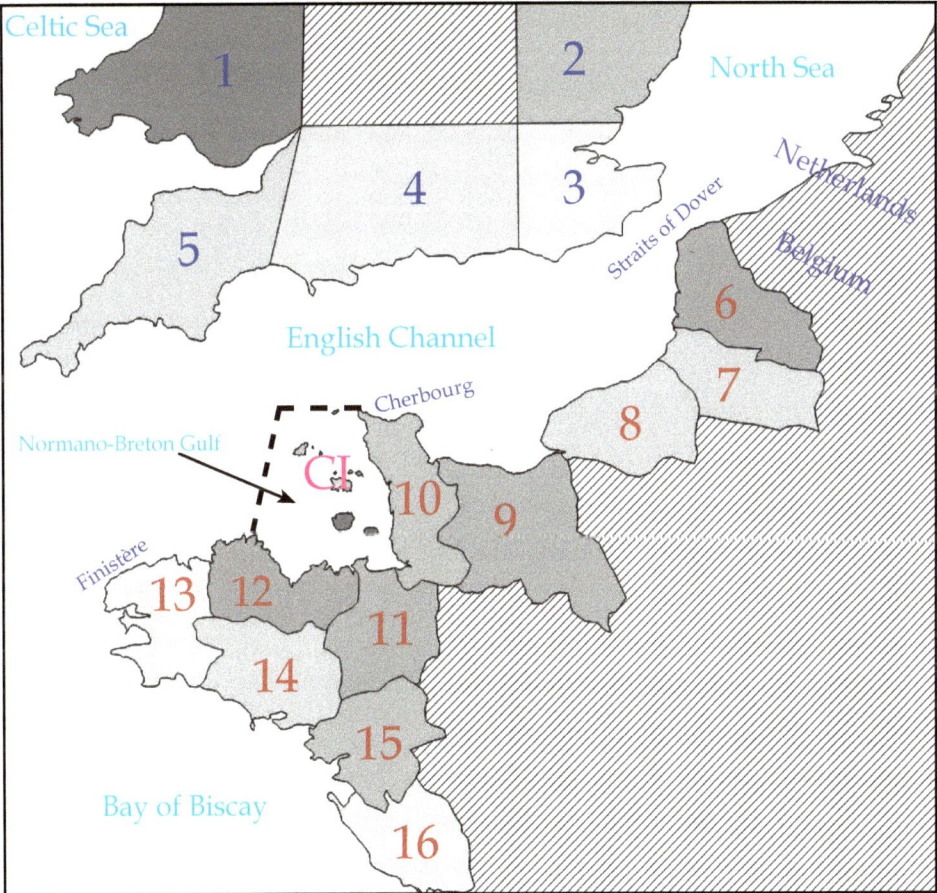

*Figure 2.3 - The geographic area from which non-native marine species are most likely to reach the Channel Islands. The numbered areas were used as reporting zones when surveying the regional distribution of species (see Chapter 2.3).*
UK: **1** = *west coast;* **2** = *east coast;* **3** = *south-east coast;* **4** = *south coast;* **5** = *south-west coast;* **FRANCE, Hauts-de-France:** **6** = *Pas-de-Calais;* **7** = *Somme;* **Normandy:** **8** = *Seine-Maritime;* **9** = *Calvados;* **10** = *La Manche;* **Brittany:** **11** = *Ille et Vilaine;* **12** = *Côtes-d'Armor;* **13** = *Finistère;* **14** = *Morbihan;* **Pays-de-la-Loire:** **15** = *Loire-Atlantique;* **16** = *Vendée;* **CI** = *Channel Islands and Normano-Breton Gulf (see Figure 2.1 for details).*

Gulf to the whole of the English Channel and the Bay of Biscay places it within a marine biological province that includes the whole of Brittany and Normandy, the eastern English Channel, and the south and south-west coast of England.

The Channel Islands are therefore geographically proximal to the English Channel, the North and Irish Seas and the Bay of Biscay. It is possible that any non-native species entering Europe at any location between the North Sea and the Iberian Peninsula could eventually reach the Channel Islands. For this reason the geographical coverage of this report is necessarily wide and includes the whole of the English Channel and the northern part of the Bay of Biscay (FOA fishing areas VIIe, VIId and VIIIa; see Figure 2.3).

## 2.2 - Information Sources: Channel Islands

The location of the Channel Islands on the southern edge of the British Isles has made them attractive places to visit by naturalists and scientists for nearly two centuries. Consequently the islands have a good historical tradition of biological recording, although much of this information is spread across hundreds of books, journals and archives and was, until recently, difficult to access.

In late Victorian times the establishment of island-based learned societies such as the Société Jersiaise and Société Guernesiaise provided a locus for local researchers and a centralised repository for natural history and other datasets. During the twentieth century a majority of the biological recording within the Channel Islands was coordinated through the Sociétés Jersiaise and Guernesiaise remaining the most important source of natural history information within the islands.

In 2005 the Société Guernesiaise and States of Guernsey founded the Guernsey Biological Records Centre (GBRC) and in 2013 the States of Jersey helped establish the Jersey Biodiversity Centre (JBC) which is housed within, and partly funded by, the Société Jersiaise. The GBRC and JBC work in cooperation and act as centralised repositories for historical and contemporary species recording across the Channel Islands although some of the smaller islands, such as Alderney, also have their own record gathering systems. The GBRC and JBC maintain over 750,000 local biological records and were a major information source for this report.

A majority of the marine species records for Jersey (including those held by the JBC) originate from the Marine Biology Section (MBS) of the Société Jersiaise who, at the time of writing, hold 95,000 records relating to over 3,200 marine species. The MBS database (known as CIMLDB) contains recent records from both its own fieldwork and that of other active organisations such as Seasearch, who are responsible for a majority of subtidal recording in the Channel Islands.

The Channel Islands have many marine biological records from c. 1860 to 1920 which reflects the Victorian passion for natural history. However, from then until the early 1980s local marine biological recording was either patchy or nonexistent. Starting in 1981, annual fieldwork by Portsmouth

Polytechnic (now a university) produced a large volume of species and habitat information for several key stretches of Jersey's coastline. In the 1990s student and government commissioned studies operated across the island, especially on Jersey's south and south-east coast.

From *circa* 2010, the MBS and Jersey Seasearch started an intensive data gathering exercise across Jersey's marine territories. This research targeted all habitats and species with the eventual aim of surveying and mapping all of Jersey's intertidal areas and a representative sample of subtidal ones. This has produced over 50,000 records and includes deliberate searches for non-native species and rapid assessments of sites that might be favourable for non-native species (see Chapter 2.2).

There is good biological data for the Bailiwick of Guernsey although there has arguably been less marine surveying in recent years than on Jersey. However, certain taxonomic groups, such as seaweeds, do have a good track record of study and the presence of taxonomic experts on Guernsey means that many non-native species have been recorded several years before they were recognised in Jersey.

During the past decade, a majority of marine recording on Alderney has occurred through the Alderney Wildlife Trust which has coordinated a series of detailed intertidal surveys across the island. This includes annual field assessments for non-native species and the sponsoring of student and other projects which include the study of non-native species (Mel

*Figure 2.4 - A member of Jersey Seasearch making a quadrat survey on the seabed at Les Écréhous in July 2014. Such surveys have helped document the presence and distribution of non-native species below the low water mark.*

Broadhurst, pers. comm.). Recording on the smaller Channel Islands has been more casual although some, such as Sark, have been surveyed in details by members of Seasearch.

The coordination and publication of data on non-native marine species within the Channel Islands has not been extensive. New finds are often listed in the Société Jersiaise *Annual Bulletin* and a summary list of non-native species from Jersey (together with a basic threat assessment) is published in the annual reports of the States of Jersey Marine Resources Section. This information is compiled annually by the MBS using species records from their database.

When compiling a list of non-native species reported from the Channel Islands, the primary datasets that were consulted were the biological records databases held by the MBS, JBC, GBRC and Seasearch. Further data were provided by the Alderney Wildlife Trust, Société Guernesiaise, and the personal records of individual researchers.

## 2.3 - Information Sources: Regional

At a regional level, marine non-native species are historically less well-documented and reporting more fragmented, than their terrestrial counterparts. Even within individual countries the monitoring of marine non-natives species may vary between regions with some areas better documented than others. This unequal spread of effort and data presents an issue when trying to gain a full understanding of the diversity and nature of non-native species within the wider marine province of which the Channel Islands are a part (Figure 2.3).

In the absence of a single comprehensive resource for the Normano-Breton Gulf (although see Goulletquer, 2016), a broad survey of potential information sources was undertaken with the aim of creating a list of non-native marine species that have the potential to reach the Channel Islands. This list was initially created by synthesising several existing regional surveys such as those of Eno *et al.* (1997), Wolff (2005), Blanchard *et al.* (2010), Godet *et al.* (2010), Minchin *et al.* (2013), Roy *et al.* (2012) and Roy *et al.* (2014).

This information was supplemented by a wider literature search using taxonomic guidebooks, regional studies, the Web of Knowledge and other sources. Species records were also sought from a range of databases including NBN Gateway, GBNNSS, JNCC, MarLin, INPN, MNMH and DAISIE. The current taxonomic status for individual species was checked using the WoRMs database.

This survey looked at over two hundred non-native marine species that have been reported from North-west Europe. Each of these was assessed to see if it had the potential to reach the Channel Islands (based on observed population behaviour) and whether its habitat preferences and environmental tolerances could allow it to establish itself in Channel Island waters. This was done to exclude any exotic species, such as tropical animals washed in on driftwood that cannot reproduce locally and any species that would be unlikely to survive due to the properties of Channel Island waters (e.g. brackish water organisms as there are no estuaries within the islands).

This assessment produced a list of 134 non-native marine species which have either already been reported from the Channel Islands or were considered to have the potential to reach the islands and establish local populations. A further information search (using mostly the same sources listed earlier plus targeted searches using academic databases) sought to obtain the following information for each shortlisted species:

- Its current taxonomic status.
- The location and dates of all Channel Island records.
- The location and dates of records from all English Channel coasts, the UK, Netherlands and Atlantic French coast.
- The location and date of the first European report(s).
- The probable transport vector into Europe.
- The probable transport vector for species with Channel Island records.
- Its ecological and habitat preferences.
- The known or suspected effect of a species on habitats, species, environmental health, economy and ecosystem functions.

This information was analysed with a view to understanding the diversity, abundance, population dynamics and potential threat exhibited by each non-native species which has the potential to be found in Channel Island waters. A list of the species assessed is given in Appendix I and the results of the analysis are presented in Chapter 3.

## 2.4 – Threat Assessment

Non-native species can present a variety of threats to the areas that they enter. These threats may be to biodiversity, ecology and environmental health but they can also affect human health and even local economies.

Recent European studies have used differing methodologies to assess and quantify the threat presented by non-native marine species. These methodologies usually focus on the central features relating to an individual species' establishment in a region. This may include the means of a species' arrival and establishment, its method of spread and its impact on local ecosystems, species, economies and health.

All such assessment schemes are necessarily subjective and the complexity of their criteria and the amount of information and techniques they utilise varies considerably. Sometimes there is little overlap between different assessment systems formulated for individual countries and projects making it difficult to apply the results from localised surveys to other regions (Verbrugge *et al.*, 2012).

When quantifying the potential threat presented by 134 non-native species listed in Appendix I, a straightforward assessment technique was desirable due to the number of species involved and the large volume of information that had been gathered on them. After investigating several recent assessment schemes, it was decided to adapt the scoring system developed by Roy *et al.* (2014a) which was itself modified from Branquart (2007). This assesses each species for five parameters, rating its potential impact from one to five. When the individual parameter scores are multiplied together they provide an overall threat score for the species.

For this report scores of between one and five were provided for four widely used parameters: ecosystem services; habitats and species; disease; and economic impact.

Two other parameters were added. One to quantify the local distribution and breeding status of those non-native species that have been reported from the Channel Island. The other to assess the likelihood of a non-native species not recorded from the islands arriving and establishing itself. Scores for each parameter were awarded following an assessment of all available information for a species. A list of the parameters and the scoring systems used is given below. The individual scores may be seen in Appendix II.

## 1 - Impact: Ecosystem Functions

The likelihood of a non-native species altering ecosystem services and functions such as nutrient cycling, physical alteration, succession and food webs. Graded using: 1 (None/Minimal); 2 (Minor); 3 (Moderate); 4 (Major); 5 (Massive).

## 2 - Impact: Habitats and Species

The likelihood of a non-native species altering local habitats and species via competition, alteration of habitats or genetic effects (introgression). Graded using: 1 (None/Minimal); 2 (Minor); 3 (Moderate); 4 (Major); 5 (Massive).

## 3 - Impact: Disease and Poisoning

The likelihood of a non-native species affecting native species or humans via disease, toxins or predation. This includes the transmission of parasites, viruses and other pathogens between species (including humans) and the production of toxins or poisons that could affect local wildlife or the higher food chain. Graded using: 1 (None/Minimal); 2 (Minor); 3 (Moderate); 4 (Major); 5 (Massive).

## 4 - Impact: Economic

The likelihood of a non-native species being able to impact local economies. Substantial effects might include mass mortality of farmed or economically important species through pathogens or parasites. Lesser effects may be caused by costs incurred through predation, grazing, fouling, effects on tourism and commercial fishing. Graded using: 1 (None/Minimal); 2 (Minor); 3 (Moderate); 4 (Major); 5 (Massive).

## Threat Score

The numerical grades produced from the above four parameters were multiplied together to produce an overall 'threat score'. Although subjective, this methodology offers a general guide to the effect that a species may be expected to have after its arrival in the Channel Islands and permits the ranking of species by their threat score.

Two further parameters were added to quantify the dispersal characteristics of each species. The first was applied to species already known to occur within the Channel Islands. The second is for those species that have the potential to reach the islands. Neither was used in the calculation of the overall threat score.

## Dispersal: Species Established in the Channel Islands

Only for those non-native species which have a contemporary or historical record from Channel Island waters or those whose regional occurrence and ecology suggests that they are already in the Channel Islands. Scoring is as follows:

1 - No known records and probably not established (assessed using Horizon Scanning; see below).
2 - Historically reported but probably extinct locally.
3 - No records but probably in CI for several years.
4 - Established but not spreading or spreading slowly.
5 - Established and spreading rapidly.

## Horizon Scanning: Species from Neighbouring Regions

Only for those non-native species in Appendix I that do not have records from the Channel Islands but have the potential to reach there. Scoring is as follows:

1 - Establishment is unlikely.
2 - Establishment possible within 20 years.
3 - Establishment possible within 10 years.
4 - Establishment possible within 5 years.
5 - Establishment imminent.

The results of the threat assessment are presented in Appendix II. The results of the survey and its analysis are discussed Chapter 3 with a list of individual species being presented in Chapters 5 (animals) and 6 (plants).

# 3 - Results and Discussion

The survey process described in Chapter Two produced a shortlist of 134 non-native marine species whose preferences and behaviour suggests that they have the potential to be found in Channel Islands territorial waters. An assessment looked at several key parameters that could assist with understanding the dynamics, biology, behaviour, ecology and effect that each of these species might have, should they become established in the Channel Islands.

This chapter will look at the results of this assessment in relation to the species shortlisted and especially those that have already been recorded from the Channel Islands area.

## 3.1 - Taxonomic Diversity

Of the 134 non-native species shortlisted in Appendix I, 43 (32%) have confirmed reports from the Channel Islands (see Appendix III). A further 49 (36%) are known from elsewhere within the Normano-Breton Gulf with the remaining 42 (31%) being reported from neighbouring areas in the English Channel, Bay of Biscay or further afield.

A taxonomic breakdown of the 134 species shows that half are either a red seaweed, mollusc or arthropod (mostly crustaceans) although other significant taxonomic groups include phytoplankton, bryozoans and ascidians (tunicates). This pattern of taxonomic diversity is similar to that observed in other regional non-native marine species lists from within North-west Europe (e.g. Eno *et al.*, 1997; Wolf, 2005; Goulletquer, 2016).

The taxonomic diversity of non-native marine species recorded from the Channel Islands is similarly dominated by red seaweeds and molluscs but there are proportionately fewer species recorded from the annelida, phytoplankton and arthropoda (Table 3.1; Figures 3.1, 3.2). It is probable that these groups have been historically under-recorded in the Channel Islands and that there are species awaiting discovery in local waters.

The identification of annelids, ascidians, red seaweeds, plankton, barnacles and small arthropods often requires expert knowledge and equipment. These species may be microscopic or live in hard to reach places, such as muddy harbours, under pontoons or offshore. This may explain why there are comparatively fewer records for these groups from the Channel Islands than the region as a whole.

The only practicable solution to the issue of understudied taxonomic groups is to get expert help from off island (via visiting experts or sending specimens for identification) or to train up local individuals to recognise and identify species. Many recent identifications of non-native marine species in the Channel Islands have been made by scientists who were invited to the islands to participate in surveys (e.g. via Seasearch) or to undertake consultancy work. This highlights the value of encouraging or paying for experts to visit the Channel Islands.

At 43, the total number of non-native marine species recorded from the Channel Islands is lower than for the wider geographical region in which they reside. This is not surprising given that most of the European species listed in Table 3.1 have yet to reach the islands. Furthermore, differences in available habitats and other ecological parameters mean that not every species listed in Appendix I will be capable of becoming established in the islands. Nonetheless, it is expected that the total number of non-native marine species known from the Channel Islands will continue to rise steadily as pre-existing plants and animals are identified and new ones are introduced into local waters.

The taxonomic composition of non-native species known from the Channel Islands against those of the wider area suggests that several phyla have been under-recorded locally. This includes phytoplankton, annelids, barnacles, sea squirts and red seaweeds, many species of which require specialist knowledge and literature in order to facilitate identification. Phyla that have a good track record of identification within the Channel Islands includes molluscs, large crustaceans and brown seaweeds, most of

| Phylum: Class | Group | All Species | CI Species |
|---|---|---|---|
| Cercozoa | Protists | 3 | 1 |
| Porifera | Sponges | 1 | 0 |
| Cnidaria: *Anthozoa* | Jellyfish | 2 | 1 |
| Cnidaria: *Hydrozoa* | Hydroids | 4 | 0 |
| Ctenophora | Sea combs | 1 | 0 |
| Platyhelminthes | Flatworms | 2 | 0 |
| Nematoda | Nematodes | 1 | 0 |
| Annelida: *Polychaeta* | Worms | 9 | 1 |
| Arthropoda: *Lower crustacea* | Barnacles; Ostracods | 16 | 2 |
| Arthropoda: *Higher crustacea* | Crabs; Prawns | 11 | 3 |
| Arthropoda: *Pycnogonida* | Sea spiders | 1 | 0 |
| Mollusca: *Bivalvia* | Clams | 10 | 5 |
| Mollusca: *Gastropoda* | Sea snails | 8 | 3 |
| Bryozoa | Sea mat | 8 | 4 |
| Chordata: *Actinopteri* | Fish | 1 | 0 |
| Chordata: *Ascidiacea* | Sea squirts | 6 | 4 |
| Myzozoa: *Dinophyceae* | Dinoflagellates | 7 | 0 |
| Ochrophyta: *Bacillariophyceae* | Diatoms | 9 | 2 |
| Ochrophyta: *Raphidophyceae* | Micro-algae | 2 | 0 |
| Ochrophyta: *Phaeophyceae* | Brown seaweeds | 3 | 3 |
| Chlorophyta | Green Seaweed | 1 | 2 |
| Rhodophyta: *Florideophyceae* | Red seaweeds | 22 | 10 |

*Table 3.1 – A taxonomic breakdown of non-native marine species from: (1) the English Channel/north Biscay geographic area; (2) the Channel Islands.*

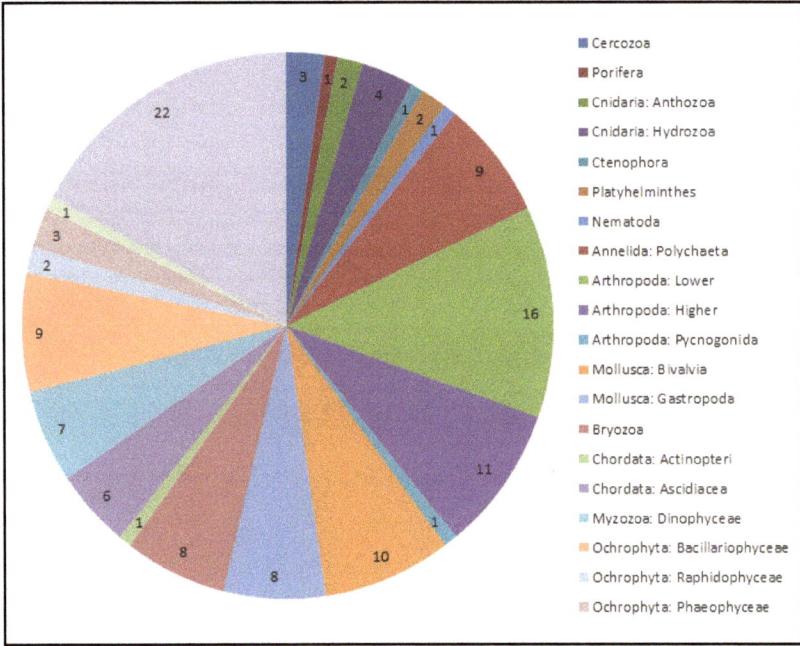

*Figure 3.1 – Taxonomic breakdown of the species listed in Table 3.1 by phylum and class. The figures refer to the number of species.*

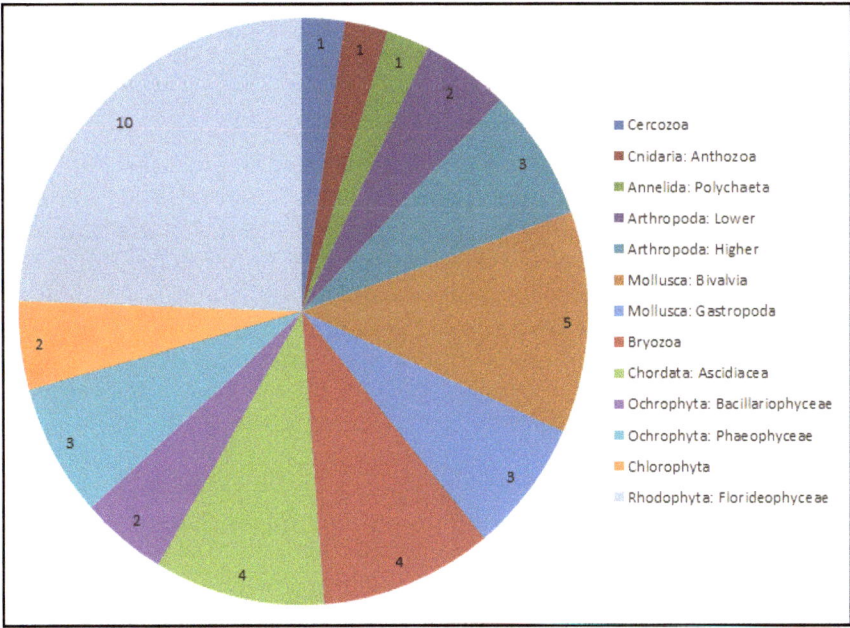

*Figure 3.2 – Taxonomic breakdown of non-native species reported from the Channel Islands by phylum and class (Table 3.1). The figures refer to the number of species.*

which are relatively easy to spot and identify by fishermen and amateur naturalists.

The difference between the taxonomic composition of species recorded in the Channel Islands and those recorded in the wider geographic province suggests that some groups of plants and animals are being under-recorded locally. These are mostly species that are traditionally hard to identify and includes members of the phytoplankton, annelids, barnacles, sea squirts and red seaweeds. This has implications for future identification and monitoring strategies; this is discussed further in Chapter Seven.


## 3.2 - Habitat Preference

Non-native marine species have habitat and environmental preferences in the same way as native species which means they will selectively colonise particular ecological niches. General environmental tolerances are known for most non-native marine species in Europe but often the specific habitats (also called biotopes) they will inhabit are not listed.

Much of the recent field data collected by the Société Jersiaise Marine Biology Section (MBS) and Jersey Seasearch not only records each species' location but also its abundance (using the SACFOR scale) and the biotope in which it has been seen (as defined by the JNCC/EUNIS classification scheme).

The MBS database (which includes a copy of the Jersey Seasearch data) holds 2,129 records of non-native species which are linked to a specific biotope. These data have been queried to produce a list of biotopes which have non-native species records together with their average abundance. This information is displayed in Appendix IV with summaries in Tables 3.2 and 3.3.

The significance of this information must be tempered as some biotopes are more extensive or accessible than others and so may have been more intensively studied. This is particularly true of intertidal biotopes for which there are many more records than those that are subtidal. Equally well, some non-native species are easier to identify in the field than others, favouring more regular recording than the smaller or hard to identify species. Nonetheless, the data in Appendix IV offer an insight into the relationship between Channel Island biotopes and non-native species.

Most biotopes have only a small number of non-natives species associated with them (Table 3.2) but 11 biotopes have five or more non-native species recorded. Similarly, Table 3.3 suggests that most non-native species have a restricted ecological tolerance and have been recorded from a small number of biotopes. However, a small number have been recorded from a large number biotopes.

The conclusion from this is that a relatively restricted number of non-native species are found across a wide range of habitats. This includes species such as *Crepidula fornicata*, *Sargassum muticum*, *Styela clava* and *Watersipora subatra*, all of which are considered to be a potential threat to the marine environment and all of which scored highly in the Channel

| JNCC Code | EUNIS code | No of Species |
|---|---|---|
| Flooded Gully Complexes | A2.872 | 14 |
| LR.FLR.Rkp.FK | A1.412 | 10 |
| Pelagic water column | | 2 |
| LR.LLR.F.Fves | A1.313 | 9 |
| LR.HLR.MusB.Sem | A1.113 | 8 |
| LR.HLR.FR.Mas | A1.125 | 8 |
| LR.MLR.BF.Fser | A1.214 | 6 |
| LR.MLR.BF.Fser | A1.214 | 6 |
| SS.SCS.ICS.MoeVen | A5.133 | 6 |
| LR.LLR.F.Asc | A1.314 | 5 |
| LS.LSa.MuSa.Lan | A2.245 | 5 |
| CR.FCR.FouFa | A4.72 | 10 |
| LR.LLR.F.Fspi | A1.312 | 4 |
| LR.FLR.Rkp.SwSed | A1.413 | 4 |
| LR.FLR.CvOv.SpR | A1.446 | 4 |
| LS.LSa.MuSa.MacAre | A2.241 | 4 |
| SS.SMp.Mrl | A5.51 | 4 |
| LR.HLR.FR.Coff | A1.112 | 3 |
| LR.FLR.Rkp.Cor | A1.411 | 3 |
| LS.LMp.LSgr.Znol | A2.6111 | 3 |
| IR.MIR.KR.Ldig | A3.211 | 3 |
| SS.SMp.SSgr.Zmar | A5.5331 | 3 |
| LR.HLR.MusB.Cht | A1.112 | 2 |
| IR.HIR.KFaR.FoR.Dic | A3.1161 | 2 |
| IR.LIR.K.Sar | A3.315 | 2 |
| CR.HCR.XFa.ByErSp | A4.131 | 2 |
| SS.SSa.IFiSa.IMoSa | A5.231 | 2 |
| SS.SMx.IMx.CreAsAn | A5.431 | 2 |
| SS.SMp.KSwSS | A5.52 | 2 |
| LR.HLR.FR.Osm | A1.126 | 1 |
| LS.LCS.Sh.BarSh | A2.111 | 1 |
| LS.LSa.MoSa.Ol | A2.222 | 1 |
| LS.LSa.FiSa.Po | A2.231 | 1 |
| LS.LSa.MuSa.CerPo | A2.242 | 1 |
| LS.LMx.Mx.CirCer | A2.421 | 1 |
| IR.MIR.KR.XFoR | A3.215 | 1 |
| IR.FIR.IFou | A3.72 | 1 |
| SS.SCS.ICS.SLan | A5.137 | 1 |
| SS.SSa.IMuSa | A5.24 | 1 |

*Table 3.2 – A list of Jersey marine biotopes which have had non-native species recorded from them. See Appendix IV for the biotope descriptions and a detailed breakdown of the species' biotope associations. (Source: Société Jersiaise)*

Islands threat assessment (Chapter 3.6). This suggests that analysing the biotope preference for individual non-native species may be a useful means of gauging their wider threat to the marine environment.

There also appear to be certain habitats that are attractive to a diverse number of non-native species. Of particular significance are intertidal biotopes that are associated with moderate to high biodiversity. This includes biotopes that retain water at low tide (e.g. harbours/marinas, rock pools and flooded gully complexes) and biotopes that have seaweed cover. It would also seem that lower shore biotopes generally have more non-native species than those on the upper and middle shore. Based on this, Channel Island non-native species seem to flourish better in sheltered,

| Species | No. of Biotopes |
|---|---|
| Crepidula fornicata | 29 |
| Sargassum muticum | 24 |
| Styela clava | 13 |
| Crassostrea gigas | 11 |
| Grateloupia subpectinata | 10 |
| Watersipora subatra | 10 |
| Codium fragile fragile | 9 |
| Undaria pinnatifida | 8 |
| Asparagopsis armata | 7 |
| Grateloupia turuturu | 4 |
| Perophora japonica | 4 |
| Tapes philippinarum | 4 |
| Botrylloides violaceus | 3 |
| Gracilaria vermiculophylla | 2 |
| Polysiphonia harveyi | 2 |
| Antithamnionella ternifolia | 1 |
| Bugula neritina | 1 |
| Bugulina stolonifera | 1 |
| Corella eumyota | 1 |
| Austrominius modestus | 1 |
| Hemigrapsus sanguineus | 1 |
| Heterosiphonia japonica | 1 |
| Monocorophium sextonae | 1 |
| Polyopes lancifolius | 1 |
| Solieria chordalis | 1 |
| Tricellaria inopinata | 1 |

Table 3.3 – A list of Jersey non-native marine species and the number of biotopes in which they have been recorded. See Appendix IV for more details. (Source: Société Jersiaise)

permanently damp, high productivity habitats suggesting that biotopes associated with higher productivity and biodiversity may be more prone to non-native species colonisation. If so, then this may be a cause for concern as it is often higher productivity biotopes that support ecosystem functions that provide significantly beneficial environmental and economic services.

The Société Jersiaise is currently engaged in a project to map in detail all the intertidal and shallow marine biotopes around Jersey (Figure 3.3). This information will eventually allow projections to be made as to the probable spread and impact of individual non-native species based on their known association with, and abundance within, particular marine biotopes (see Chambers, Binney and Jeffreys, 2016).

The number of biotopes within which a non-native species has been reported offers an insight into their range of environmental tolerance. This is important because those species that have a wide environmental tolerance can be more problematic than those that can only survive in a narrow range of conditions.

Regular monitoring of the number of biotopes in which a non-native species has been reported will help to gauge its spread into the wider environment and as such act as an early warning system with regard to its overall threat to local habitats and species.

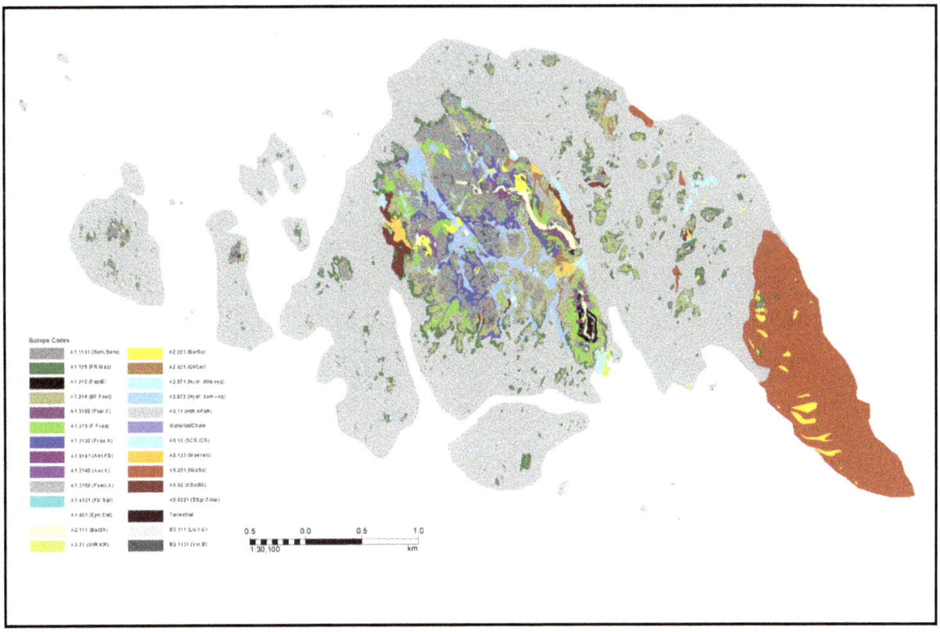

*Figure 3.3 - A biotope map for Les Écréhous, a reef located 10 km north-east of Jersey. Detailed habitat maps such as this are important for documenting and predicting the spread of non-native species.*

## 3.3 – Date of First Observation

The date on which a non-native species is first recorded is of historical interest but it has a scientific significance too. Knowing when a species entered a region can give an indication of the method of introduction and the rate and means of spread following its establishment.

In most instances a non-native species will have been present for several years or decades before it is first identified although visually obvious ones, such as *Sargassum muticum*, may be reported only a short time after establishment (see Bracken, 2012). It is therefore assumed that the date of first observation lags some time behind the actual date of introduction and that the place of first observation may not necessarily be the actual point of arrival. This is particularly true for hard to identify and microscopic species which may go unrecorded for many decades.

Using data from the survey described in Chapter Two, Figure 3.4 shows the decade in which the 134 non-native marine species listed in Appendix I were first recorded across three overlapping regions:

1 - North-west Europe from the Bay of Biscay to the southern North Sea.
2 - The English Channel and southern Brittany (Fig. 2.3).
3 - Channel Islands' territorial waters (Fig. 2.1).

The historic observation pattern for North-west Europe and the English Channel shows that the recording of non-native marine species increased markedly after the 1950s. There is a noticeable spike towards the end of the twentieth century followed by a decrease in the opening decades of the twenty-first century. This uneven pattern of recording has been previously observed and commented on and is usually interpreted as being due to an increased rate of species' introduction following World War II (see Wolff, 2005; Minchin *et al.*, 2013).

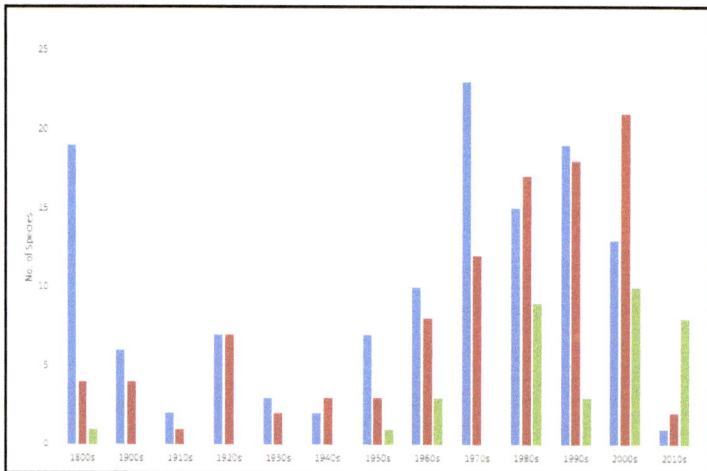

*Figure 3.4 – The decade in which the individual non-native species from Appendix I were first observed. Blue = first observation in NW Europe; red = first observation in the western English Channel/Brittany; green = first observation in the Channel Islands.*

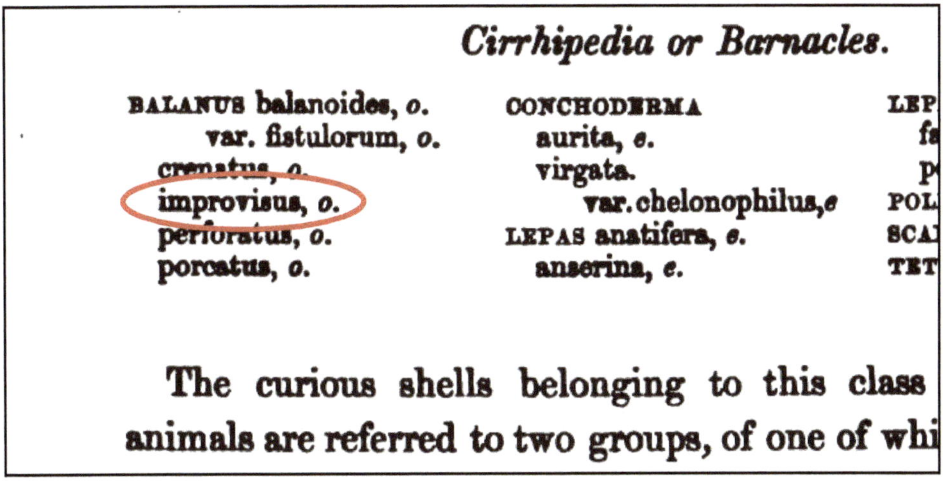

> **Cirrhipedia or Barnacles.**
>
> | | | |
> |---|---|---|
> | BALANUS balanoides, *o.* | CONCHODERMA | LEP |
> | var. fistulorum, *o.* | aurita, *e.* | fa |
> | crenatus, *o.* | virgata. | P |
> | improvisus, *o.* | var. chelonophilus,*e* | POL |
> | perforatus, *o.* | LEPAS anatifera, *e.* | SCA |
> | porcatus, *o.* | anserina, *e.* | TET |
>
> **The curious shells belonging to this class**
> **animals are referred to two groups, of one of whi**

*Figure 3.5 - The earliest printed reference to a Channel Island non-native marine species, the barnacle* Amphibalanus improvisus, *in Ansted and Latham's 1862 volume* The Channel Islands.

The pattern of first observation in the Channel Islands differs from this. Few non-native species were recorded in the islands prior to 1980 followed by a drop in the 1990s and then a rapid increase in reports after 2000. It would seem that prior to the twenty-first century, easy to identify species, such as *Sargassum muticum* and *Crepidula fornicata*, were recorded shortly after their arrival in the Channel Islands but that smaller species and those that do not directly affect human activity, went unrecorded. This suggests that for many decades there was a considerable time lag between the arrival of a species into the Channel Islands and their being officially recorded.

This punctuated pattern of observation is probably due to biological recording within the islands being in the hands of a small group of local naturalists and visiting specialists. While such individuals are active observations are made; but when they are inactive or absent, gaps of several years may occur during which no records are made.

For example, the identification of several encrusting non-native species from Guernsey's marinas after 2000 arose from the work of one local person. Similarly, a cluster of observations in Jersey in the early 1980s arose from annual fieldwork by Portsmouth Polytechnic while another cluster of records after 2010 followed the establishment of a more formalised reporting network by the Marine Biology Section (Société Jersiaise). Many of the species identified after 2000 had probably been present locally for several years before being officially recorded.

Identifying many non-native species (and especially microscopic ones) requires specialist skills and periods of time when there is a lack of resident or visiting taxonomists will make it difficult to monitor existing non-native species and identify new ones. Until recently a lack of any formalised recording mechanisms for non-native species within NGOs and government bodies also made it difficult to maintain and coordinate records and species lists. Such issues affect many local coastal locations and explain why globally the first observation of many non-native species are from locations that are adjacent to marine laboratories or population centres with university facilities.

It is also only since the advent of the internet and a network of UK biological records centres that the coordination and publication of records has become rapid and universally accessible. Within the Channel Islands the establishment of Guernsey Biological Records Centre, the Jersey Biodiversity Centre and the Alderney Wildlife Trust has been of great assistance when it comes to recording and monitoring non-native species (terrestrial and marine) and it is envisaged that these organisations will play an important role in future monitoring schemes (see Chapter 6.3).

## 3.4 – Region of Origin

Figure 3.6 shows the global region of origin for the 134 non-native species listed in Appendix I both for the Channel Islands and their wider geographic region. The local and regional patterns are similar both to each other and to other European studies of non-native marine species. Immediately obvious is that the most significant region as a source of non-native marine species is the Pacific Ocean particularly its north-western sector where countries like Japan and Korea have oceanographic conditions that are similar to those in north-western Europe.

In fact, there has been a recent historical shift in the region of origin as pre-World War II many non-native marine species came from the western Atlantic Ocean. After the War this changed to the Pacific Ocean, something that probably reflects the dramatic increase in commercial shipping and aquaculture seed stock moving between the Pacific and Europe.

For reasons that are set out in Chapter 1.2, the principal methods of introduction for non-native marine species (shipping and aquaculture) means that there are locations within Europe which are more liable to receive new arrivals than others. Such 'hubs' include large ports receiving international shipping and areas of extensive or intensive aquaculture, particularly of shellfish.

Figure 3.7 shows the first recorded regional location for all the non-native marine species in Appendix I and, from this same list, those species that have been reported from the Channel Islands. This gives a crude indication as to which areas of Europe might be supplying non-native marine species to the Channel Islands, something that is useful to know when trying to predict the likely threat from future arrivals elsewhere in the region.

These data suggest that a majority of non-native species currently known from the Channel Islands originally arrived in Europe to the north or east of the islands in the English Channel or North Sea. These areas are known to have a heavy concentration of non-native species which probably reflects the concentration of ports, international shipping and aquaculture centres along these coasts.

Channel Island non-native species which originated to the south (i.e. Brittany and the Bay of Biscay) almost all entered the area through the aquaculture industry. This may be a cause for concern as species that arrived in southern Brittany and Biscay currently form a minority of the known non-natives in the Channel Islands. However, their regional occurrence and pattern of spread suggests that in future years the islands could be receiving an increased number of species originating from the south (i.e. the Brittany coast).

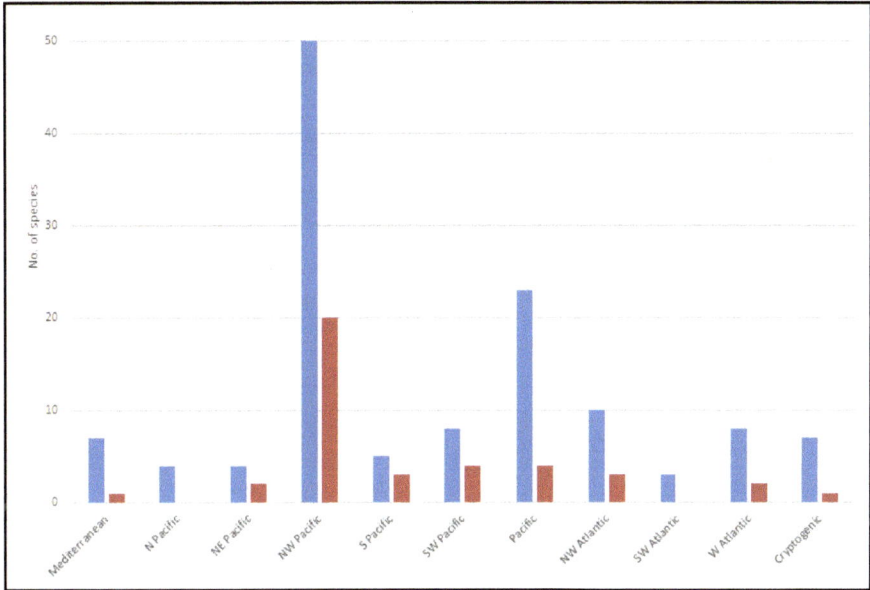

*Figure 3.6 – The region of origin for the individual non-native species from Appendix I. Blue = all species; red = species recorded from the Channel Islands. See Appendix I for more detail.*

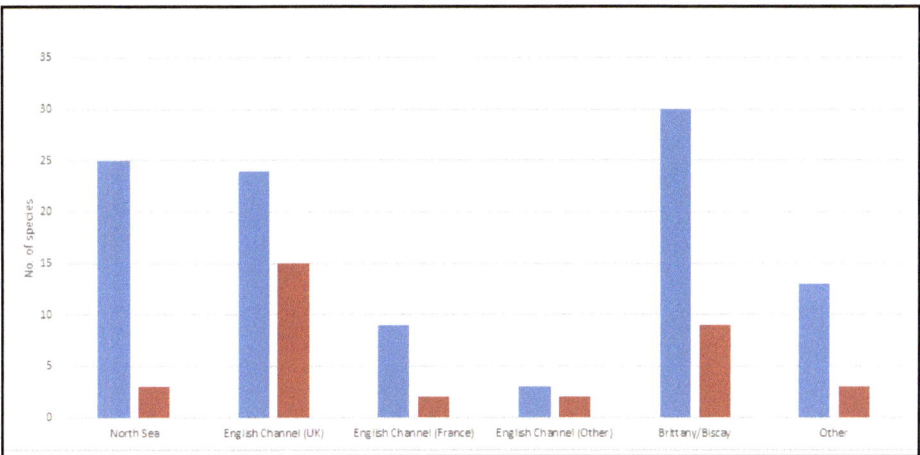

*Figure 3.7 – The first reported regional location for all non-native species listed in Appendix I (blue) and those recorded from the Channel Islands (red).*

## 3.5 – Transport Vectors

The principal transport vectors for the non-native species listed in Appendix I from their region of origin to Europe are shipping (tank ballast and hull fouling) and aquaculture (usually organisms that live on or inside farmed species or the farmed species themselves). The balance and range of vectors for the English Channel and northern Biscay region (Figure 3.8) resembles that given for European non-native species in other studies.

Figure 3.9 shows the probable transport vector for non-native marine species that have been recorded from the islands. This differs considerably from Figure 3.8 with a majority of species (54%) most likely reaching the islands after being naturally dispersed from neighbouring areas. Given its lack of global trade links and regulated aquaculture sector, it is probable that all non-native marine species recorded from the Channel Islands arrived from other parts of Europe and not directly from their native region of origin.

After natural dispersal, the second most important transport vector for Channel Island species is shipping. This seems to have brought species directly into the islands' harbours and marinas from where some have entered into the wider marine environment (e.g. *Watersipora subatra*).

Jersey and Guernsey both have important commercial ports which receive ferries, cargo boats and leisure craft from France, the UK and further afield. Twelve species have their first recorded locations within the harbours at St Helier and St Peter Port suggesting that these were their point of entry (see Ryland *et al.*, 2009).

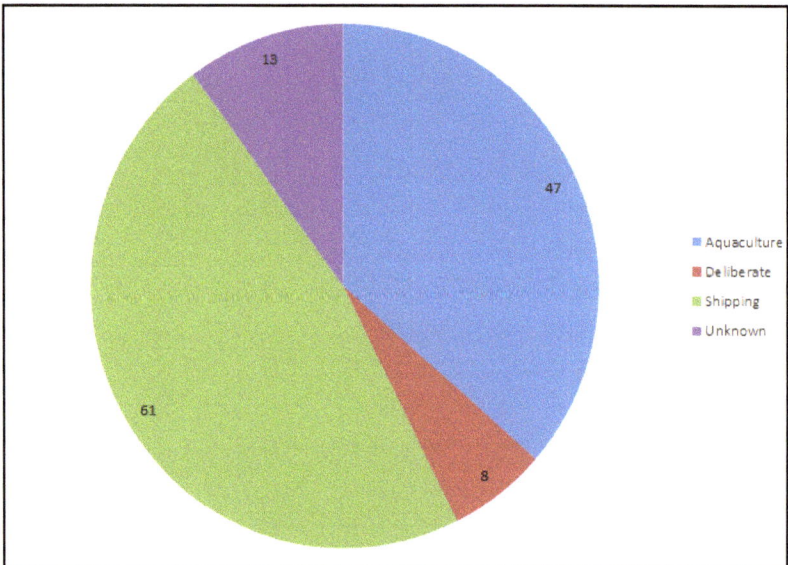

*Figure 3.8 – The vector by which the species listed in Appendix I were probably transported to Europe from their region of origin. See Appendix I for more details.*

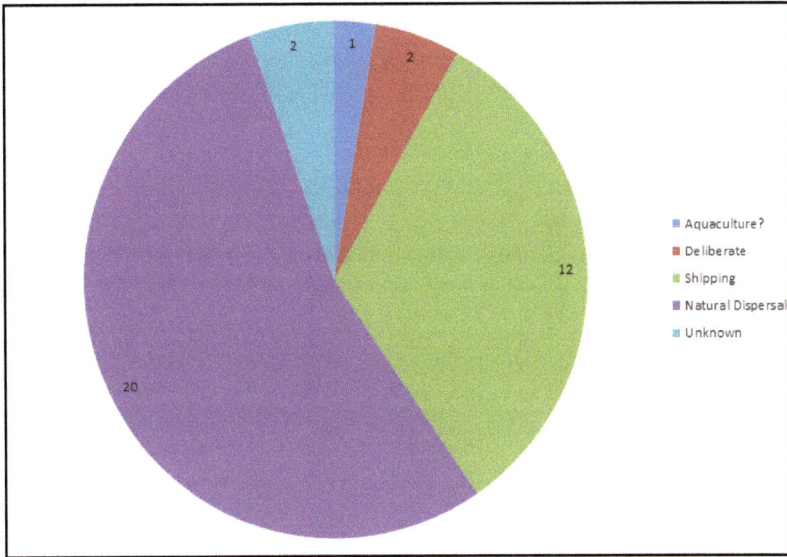

*Figure 3.9 – The vector by which non-native species recorded from the Channel Islands probably arrived in the islands from elsewhere in Europe.*

Outside of the main ports, Jersey and Guernsey also have small local harbours located around the coast, as do some of the lesser Channel Islands. These receive inter-island and regional boats but available evidence suggests that these smaller (often drying) harbours have not been colonised by non-native species to the same extent as St Helier and St Peter Port Harbours. It seems probable that the islands' marinas are particularly susceptible to non-native organisms in ways that smaller harbours are not (see Chapter 4.1).

Although Jersey has a long-standing aquaculture sector, it is only during the past decade that the industry has started to expand significantly. As a relative newcomer into the European aquaculture market, the Channel Island aquaculture industry has benefitted from strict biosecurity measures that did not exist historically. This may have protected the island from a diverse range of non-natives that have travelled with spat/seed across Europe between other aquaculture regions.

Just two deliberately introduced aquaculture species seem to have established themselves in the wild: *Ruditapes philippinarum* and *Crassostrea gigas*. Just one species, *Polyopes lancifolius*, could have been introduced accidentally via aquaculture although this is by no means proven.

*Figure 3.10 - Oyster trestles on Jersey's east coast. Unlike many other European areas, Channel Islands aquaculture is probably not a significant transport vector.*

45

## 3.6 - Threat Scores and Horizon Scanning

The formulation of individual threat scores as outlined in Chapter 2.4 helps to contextualise the individual and collective threat presented to the Channel Islands by non-native marine species. The scoring process is subjective which means the significance of the results is open for discussion but nonetheless this tool does have a role to play in awareness raising and the focusing of resources.

The assignment of a composite threat score to each non-native marine species in Appendix I produced scores that range from figures of one to 125. A majority of species (78 or 58%) have a score of less than ten and, based on present knowledge, these are considered to present a minimal threat to the Channel Islands' marine environment.

A further 21 species (15%) have scores between ten and 20 which usually means that one of more of the four individually assessed threat criteria (see Chapter 2.4) has a medium to high score. These species are considered to present a minor to moderate threat to specific aspects of the marine environment although such effects may be localised (e.g. fouling in harbours).

Species with a score greater than 20 probably present a more serious threat to two or more aspects of the marine environment including the ability to alter habitats, affect local industries and even human health. A total of 33 species (24%) have scores of 20 or above and of these 13 (9%) have a score of above 30.

A small number of species scored highly in all four of the threat criteria suggesting that they may have a serious or even devastating effect on the Channel Islands marine environment. Some of these species are already resident (e.g. *Crepidula fornicata* and *Crassostrea gigas*) but others, such as *Didemnum vexillum*, have yet to be recorded.

The highest scoring species may require further investigation and for those that present the most serious threat, the formulation of action plans may be required to monitor and manage, if possible, their effect on the local marine environment.

The two dispersal criteria in Chapter 2.4 have been used to evaluate the rate of spread of non-native species in the Channel Islands and the wider geographic region. Of the 134 species evaluated, 43 (32%) have already been recorded from Channel Island waters or are thought to be present but as yet unrecorded. Of these, 14 species have a threat score of 20 or more including the two highest scoring species, *Sargassum muticum* and *Crepidula fornicata*.

Among those species that are not thought to be present in the Channel Islands, the rate and direction of spread suggest that 27 could reach the islands and become established within 20 years (i.e. by 2037); that nine might be established within ten years (by 2027); and eight within five years (2022). The establishment of five species is considered to be imminent and they could be recorded at any time.

There are many reasons to conduct 'horizon scanning' for the arrival of non-native species but its principal value is in the ability to be prepared and to implement monitoring and preventative measures such as biosecurity. For example, the Jersey Marine Resources Section has been regularly checking St Helier harbour for the ascidian *Didemnum vexillum*, a fouling organism that, if caught early enough, can be eradicated. If caught too late, its removal can cost many thousands of pounds.

Within the Channel Islands there is a differential rate of spread among known non-native species. Of the 43 species recorded from the islands, three are known from historical reports and are probably locally extinct. A further 29 species have recent records but are thought to be static or spreading slowly. Eight species are spreading and it is these that are of most concern at present, especially as four of them were first recorded during the last decade. An additional 25 species have a distribution and other characteristics which suggest that they may already be established in the islands but that they have not yet been formally recorded.

To be forewarned is to be forearmed and the scores from the threat and distribution assessments can be used to establish which species need to be monitored and which islands should expect to receive in the near future. Threat scores can also play a role in awareness raising as they offer an easy means of ranking species by their probable effect on the local marine environment.

| Species Name | Taxonomic Group | Threat Score |
|---|---|---|
| *Crepidula fornicata* | Mollusca Gastropoda | 125 |
| *Sargassum muticum* | Seaweed (Brown) | 100 |
| *Rapana venosa* | Mollusca Gastropoda | 80 |
| *Didemnum vexillum* | Chordata Tunicata | 80 |
| *Crassostrea gigas* | Mollusca Bivalvia | 72 |
| *Pseudodactylogyrus anguillae* | Plathelminthes | 64 |
| *Mnemiopsis leidyi* | Ctenophora | 60 |
| *Heterosigma akashiwo* | Ochrophyta | 48 |
| *Undaria pinnatifida* | Seaweed (Brown) | 36 |
| *Schizoporella japonica* | Bryozoa | 36 |
| *Hemigrapsus sanguineus* | Crustacea Malocostraca | 30 |
| *Homarus americanus* | Crustacea Malocostraca | 30 |
| *Pachygrapsus marmoratus* | Crustacea Malocostraca | 30 |
| *Bugula neritina* | Bryozoa | 27 |
| *Bugulina stolonifera* | Bryozoa | 27 |
| *Corella eumyota* | Chordata Tunicata | 27 |
| *Styela clava* | Chordata Tunicata | 27 |
| *Schizoporella errata* | Bryozoa | 27 |
| *Megabalanus tintinnabulum* | Crustacea Maxillopoda | 27 |
| *Megabalanus coccopoma* | Crustacea Maxillopoda | 27 |
| *Alexandrium minutum* | Protozoa Myzozoa | 24 |

| *Watersipora subatra* | Bryozoa | 24 |
|---|---|---|
| *Koinostylochus ostreophagus* | Plathelminthes | 24 |
| *Hemigrapsus takanoi* | Crustacea Malocostraca | 24 |
| *Pseudomyicola spinosus* | Crustacea Maxillopoda | 24 |
| *Asparagopsis armata* | Seaweed (Red) | 24 |
| *Grateloupia turuturu* | Seaweed (Red) | 24 |
| *Grateloupia subpectinata* | Seaweed (Red) | 24 |
| *Mytilicola orientalis* | Crustacea Maxillopoda | 24 |
| *Ocenebra inornata* | Mollusca Gastropoda | 24 |
| *Bonamia ostreae* | Protozoa Haplosporidia | 20 |
| *Celtodoryx ciocalyptoides* | Porifera | 20 |
| *Tricellaria inopinata* | Bryozoa | 18 |
| *Molgula manhattensis* | Chordata Tunicata | 18 |
| *Perophora japonica* | Chordata Tunicata | 18 |
| *Caulacanthus ustulatus* | Seaweed (Red) | 18 |
| *Marteilia refringens* | Protozoa Incertae Sedis | 16 |
| *Myicola ostreae* | Crustacea Maxillopoda | 16 |
| *Mytilicola intestinalis* | Crustacea Maxillopoda | 16 |
| *Dasysiphonia japonica* | Seaweed (Red) | 16 |
| *Ammothea hilgendorfi* | Chelicerata Pycnogonida | 16 |
| *Anguillicoloides crassus* | Nematoda | 15 |
| *Gonionemus vertens* | Cnidaria Hydrozoa | 12 |
| *Ensis directus* | Mollusca Bivalvia | 12 |
| *Limnoria quadripunctata* | Crustacea Malocostraca | 12 |
| *Limnoria tripunctata* | Crustacea Malocostraca | 12 |
| *Codium fragile fragile* | Seaweed (Green) | 12 |
| *Solieria chordalis* | Seaweed (Red) | 12 |
| *Caulerpa taxifolia* | Seaweed (Green) | 12 |
| *Coscinodiscus wailesii* | Bacillariophyta | 12 |
| *Desdemona ornata* | Annelida Polychaeta | 12 |
| *Grandidierella japonica* | Crustacea Malocostraca | 12 |

*Table 3.4 – Ranked non-native species with a threat score above ten. * = species has been reported from Channel Island waters. See Appendix II for more detail.*

48

# - Part Three -

## Non-native Marine Species
### - *A List of Species* -

# 4 - Non-native Marine Species
## - *Part One: Animals* -

The following two chapters present a compendium of all 134 non-native marine species listed in Appendix I. Each species has its own entry which contains a summary of key information relating to its origin, transport vector, threat and regional/local distribution.

Most species are accompanied by a regional map which shows their distribution in relation to the Channel Islands. These maps have been created using data gathered during the survey of databases, literature and other resources outlined in Chapter Two. However, the patchy nature of coastal recording means that the actual distribution for many species will probably be different to that shown in the maps.

Those species that have been recorded around Jersey have a separate distribution map plotted out on a one kilometre grid. This preferential mapping of Jersey records is partly because this report was prepared by the States of Jersey but it also reflects the nature of marine biological surveying on Jersey which has been intensive in recent years (see Chapter 2.2).

Where possible an illustration for each species has been provided. These are intended to help the reader gain an idea as to the morphology of the species concerned just in case specimens are encountered in the field. In most instances the illustrations are not sufficient to permit a firm identification and so it is recommended that anyone wishing to confirm the identification of a specimen should check other resources or contact an expert. No taxonomic description is provided as, in most cases, original specimens have not been examined. Those seeking a more comprehensive guide to most of the species listed here should consider consulting the recent work by Philippe Goulletquer *Guides des Organisms Exotiques Marins* (Berlin, 2016).

The authors are grateful to the many people that provided illustrations for this work, either directly or via Creative Commons licencing. An attempt has been made to verify the identification of the organisms illustrated but the possibility of errors remains.

By the time of this report's publication it is probable that some of the species' information it contains will already be out-of-date. The world of non-native marine species is fast moving and keeping pace with changes in distribution, threats and management strategies is difficult. Do not assume that any of the information in the pages that follow is the latest word on a species' characteristics or behaviour. New discoveries are constantly being made and further checking is recommended.

NOTE: At the final stages of preparing this report the authors became aware of four additional non-native species. These have been added to this section but are not included in statistics quoted elsewhere in the report. The species are: *Borylloides diegensis, Chrysymenia wrightii; Cryptonemia hibernica; Dictyota cyanoloma.*

# Key to Species Entries

Identification may require expertise

Individual threat scores

Taxonomic group/ common name

Species name

Overall threat score

## Bugula neritina
### Bryozoan

Threat score: 27

| ES | Hab | Tox | Econ |
|----|-----|-----|------|
| 3 | 3 | 1 | 3 |

Regional distribution

Jersey distribution

Habitat preference (Jersey only)

Habitat: Artificial structures within marinas.

The history of this species in Europe is complex. First identified in the 1911 in Plymouth, *Bugula neritina* was afterwards reported from southern England and France but was for a short while in the 1990s declared extinct in the UK. It UK presence was re-established in 2004 and the first Channel Island report came from Guernsey marina in 2007. It was probably also established in Jersey's marinas at this time although this was not confirmed until 2014. Although it continues to spread in Europe, *B. neritina* primarily remains a species that is associated with ports and marinas where it has sometimes been reported as a fouling organism. This has not been observed in Jersey where and it is considered to be a low level threat to the island's wider marine environment.

Commentary

Left: *A specimen of* B. neritina *taken from St Helier Marina in 2016 magnified x20.* Right: *Details of the zooids on the same specimen (x45).*

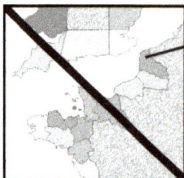

Illustrations

Horizon scanning. Species has not yet been recorded from the region.

# *Marteilia refringens*
## Incertae Sedis

| ES | Hab | Tox | Econ |
|----|-----|-----|------|
| 1  | 1   | 4   | 4    |

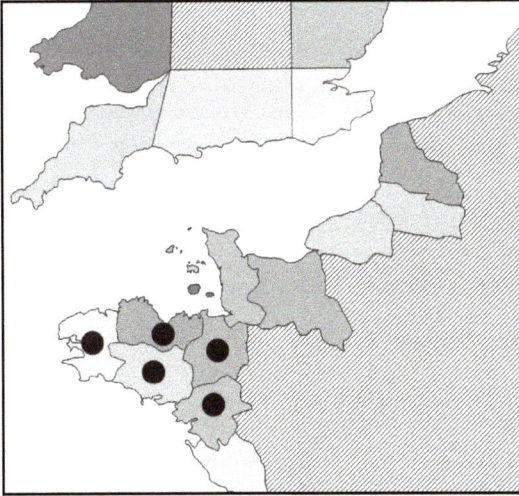

*Marteilia refringens* is a protozoan of cryptogenic origin that is associated with the lethal Marteiliosis (Aber) disease in the Flat Oyster (*Ostrea edulis*). It has been recorded from the UK to Morocco including northern Biscay where it caused problems in the Golfe de Morbihan in the 1970s. It is primarily associated with estuarine and inshore waters and is not thought to be common in higher salinity open waters. Jersey has restrictions on importing seed stock from designated areas for *M. refringens* and this organism is notifiable but has not been recorded in the Channel Islands.

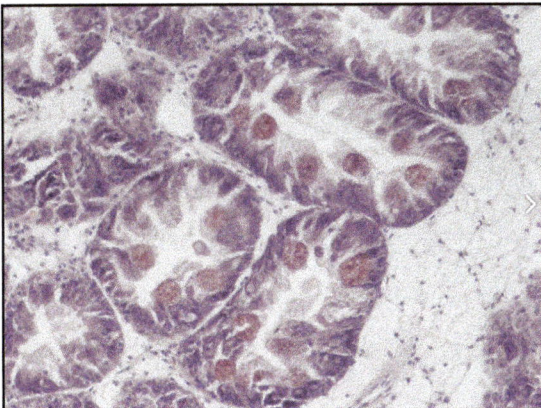

*Above:* Marteilia refringens *in the digestive gland of an oyster.*

53

# *Bonamia ostreae*
## Haplosporidia

Threat score: 20

| ES | Hab | Tox | Econ |
|----|-----|-----|------|
| 1 | 1 | 4 | 5 |

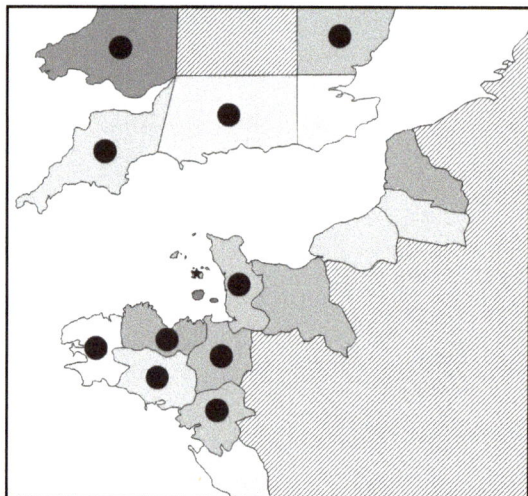

Originally from the North-west Pacific, *Bonamia ostreae* is known from all parts of the Normandy and Brittany coast. It is associated with a lethal disease affecting the Flat Oyster (*Ostrea edulis*) and was responsible for mass mortality events in Brittany in 1978 which seriously affected aquaculture production. The pathogen was reported from Jersey in aquaculture areas during the late 1970s and 1980s.

Surveillance, biosecurity and the breeding of pathogen resistant oysters has been used to address the threat produced by *B. ostreae*. Jersey has restrictions on importing seed stock from areas designated as vulnerbale to *Bonamia ostreae*. This organism is notifiable and was last reported on island in 1988.

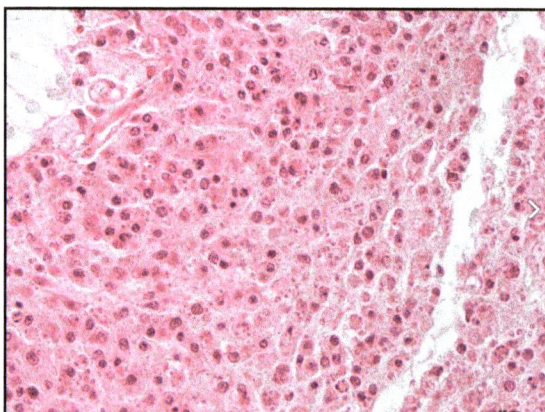

*Above:* Bonamia ostreae *infection disseminated throughout the tissue of an oyster.*

# *Haplosporidium nelsoni*
## Haplosporidia

Threat score: 9

| ES | Hab | Tox | Econ |
|----|-----|-----|------|
| 1 | 1 | 3 | 3 |

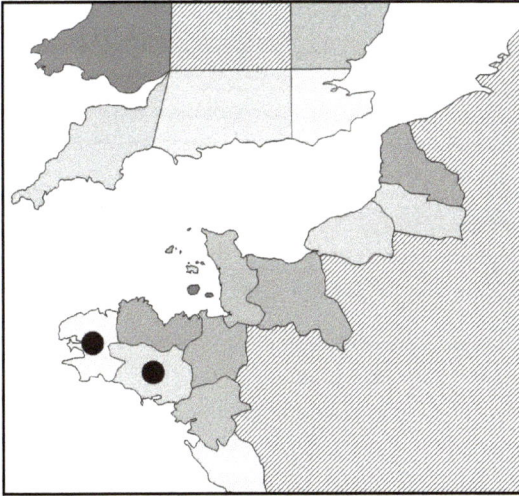

Native to the Pacific Ocean, *Haplosporidium nelsoni* is a pathogen of the Pacific Oyster (*Crassostrea gigas*) that was probably present in the region from the 1970s onwards but which was first recorded in Brittany in 1993.

This species has been associated with mass mortality events in the USA but there seems to be no record of similar events within the European aquaculture industry. Its status in the Channel Islands is unknown but at present it is not thought to present a threat to the regional aquaculture industry.

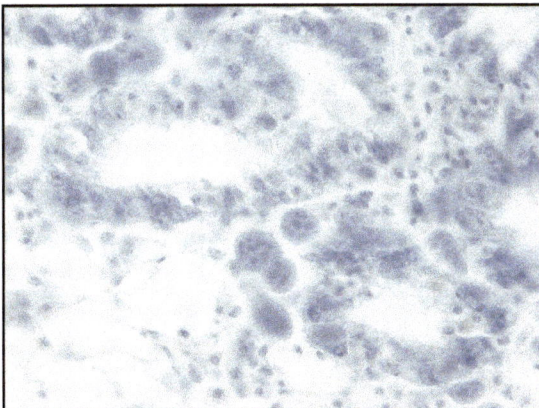

*Above: Multinucleate plasmodium stages of* Haplosporidium nelsoni. *Image width = 200 microns.*

# Celtodoryx ciocalyptoides
## Sponge

Threat score: 20

| ES | Hab | Tox | Econ |
|----|-----|-----|------|
| 5  | 4   | 1   | 1    |

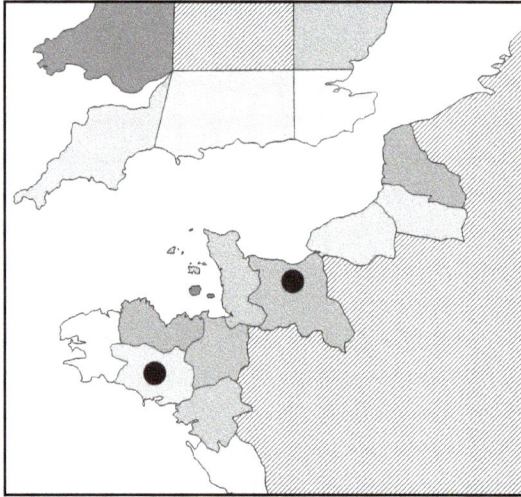

A sponge from the North-west Pacific which was reported from southern Brittany in 1999 and, in 2016, from the Netherlands and Le Havre on the English Channel. *C. ciocalyptoides* can grow on a variety of habitats and has been observed to cover sizeable areas of subtidal rock outcompeting native encrusting organisms. The ability of this species to colonise Pink Sea Fans (*Eunicella verrucosa*) is of serious concern as these are delicate and slow growing.

In southern Brittany and the Netherlands *C. ciocalyptoides* has been observed to dominate some subtidal areas and it is considered to be a serious threat to local biodiversity. Assuming its range continues to spread outwards from southern Brittany and the eastern English Channel, then *C. ciocalyptoides* will probably reach the Channel Islands. Awareness amongst local divers represents the best opportunity for monitoring as well as keeping up-to-date on its spread within the local region.

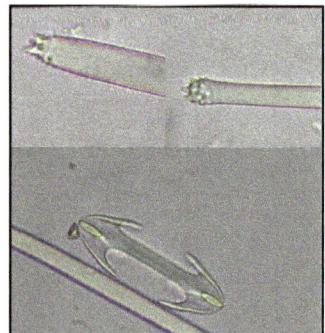

*Left: C. ciocalyptoides in situ. Right: Spicules dissolved from the sponge and viewed under the microscope.*

# *Nemopsis bachei*
## Hydroid

Threat score: 4

| ES | Hab | Tox | Econ |
|----|-----|-----|------|
| 2  | 2   | 1   | 1    |

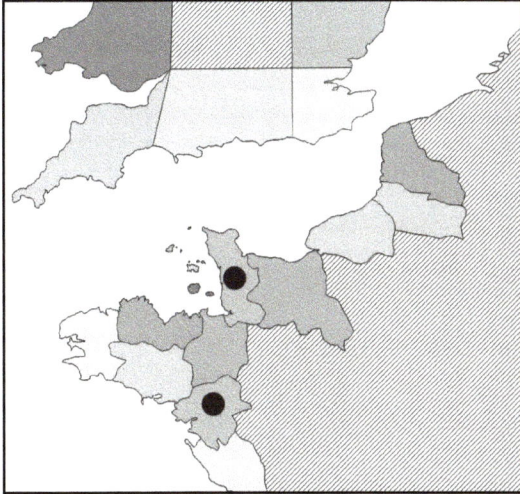

A solitary hydroid (with a medusa phase, illustrated below) from the Atlantic coast of America, *N. bachei* has been reported in Europe since at least the 1950s. Records are sporadic from Norway to southern Brittany although it is apparently established in the Netherlands. Its occurrence on the adjacent Normandy coast makes it probable that *N. bachei* is either already in Channel Island waters or will be found here eventually. It has not been considered a serious threat to local species and habitats elsewhere in Europe.

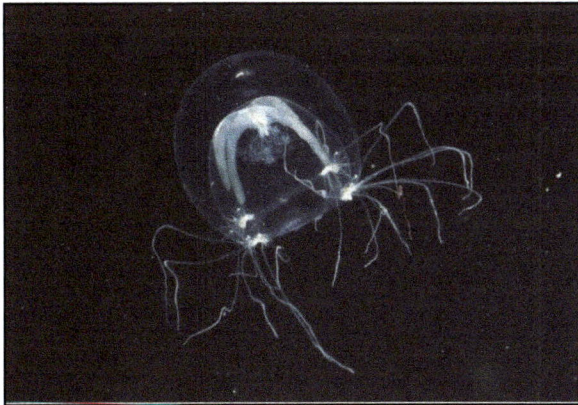

# *Gonionemus vertens*
## Clinging Jellyfish - hydroid

Threat score: 12

| ES | Hab | Tox | Econ |
|----|-----|-----|------|
| 2 | 2 | 3 | 1 |

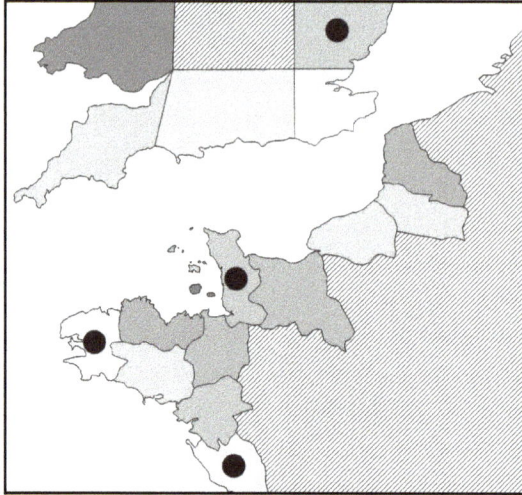

A small hydroid from the western Pacific that was introduced into the North Sea in 1913. It has a medusa stage (illustrated below) which attaches itself to algae and objects and can produce a sting that generates muscle cramps and chest tightness in humans. It has been reported from Normandy, Brittany and the Bay of Biscay but appears to be rare and of little actual threat.

Given its occurrence on the adjacent French coast, it is probably present in Channel Islands waters but its habitat, rarity and size will probbaly require a specialist to find and identify it.

# *Cordylophora caspia*
## Hydroid

Threat score: 4

| ES | Hab | Tox | Econ |
|----|-----|-----|------|
| 2  | 2   | 1   | 1    |

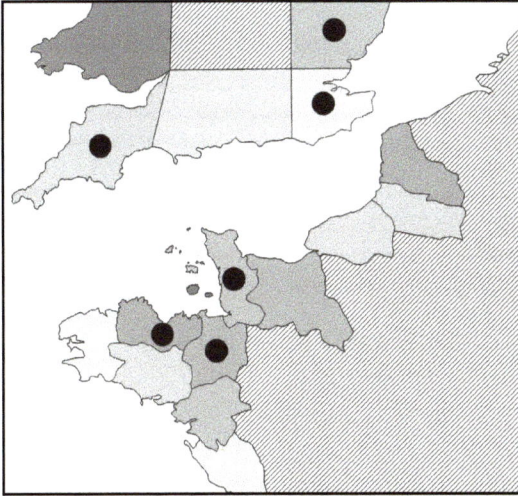

Introduced from Middle Asia in the 1930s, *C. caspia* is known from the UK, Normandy and northern Brittany. It is primarily associated with brackish waters although it may occasionally be found in fully marine conditions. It is generally unsuited to the marine areas around the Channel Islands and is unlikely to become established or have a measureable threat there.

# *Blackfordia virginica*
## Hydroid

Threat score: 4

| ES | Hab | Tox | Econ |
|----|-----|-----|------|
| 2  | 2   | 1   | 1    |

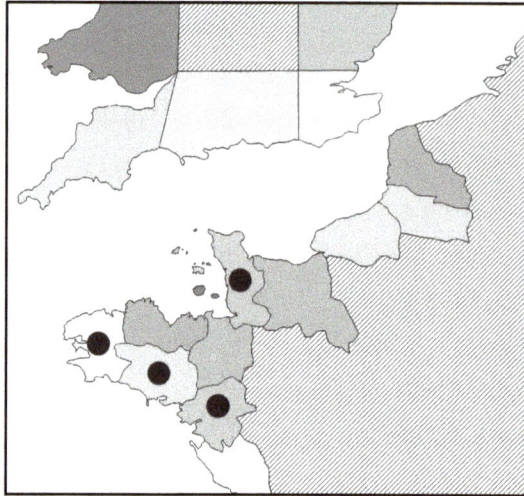

A hydroid which may be native to the Black Sea but which has been reported from around the globe. It has been present in Brittany since at least the 1950s but has only sporadic reports. The hydroid is generally associated with lower salinity conditions but the medusae phase (illustrated below) can be found in fully marine coastal areas. There are no reports from the Channel Islands but the species is known from the adjacent French coast so it is possible that it may be found locally. It is not generally considered to be a serious threat.

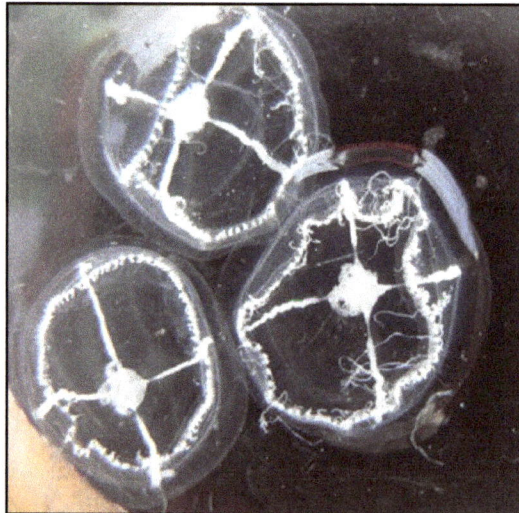

# *Diadumene cincta*
## Orange Anemone

Threat score: 4

| ES | Hab | Tox | Econ |
|----|-----|-----|------|
| 2 | 2 | 1 | 1 |

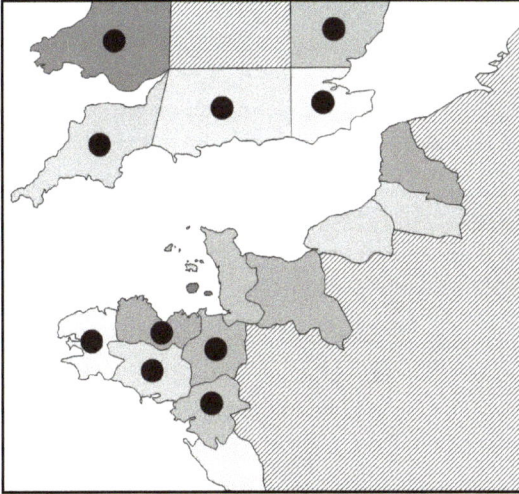

A small anemone from the North-west Pacific that has spread widely across the globe via shipping. It has been in Brittany since the 1960s and has many coastal records from the UK although none could be found for Normandy. In the English Channel *D. cincta* is primarily associated with marinas and harbours where it may be abundant. It is noted as a fouling species, particularly on pontoons and boats' hulls.

Although it can tolerate fully marine conditions, it is possible that *D. cincta* has a preference for lower salinity conditions which may lessen the probability of its establishment in the Channel Islands. Given its current rate of spread and distribution it is possible that this species will be found locally, even if it does not become common.

# *Diadumene lineata*
## Orange-striped Anemone

Threat score: 4

| ES | Hab | Tox | Econ |
|----|-----|-----|------|
| 2 | 2 | 1 | 1 |

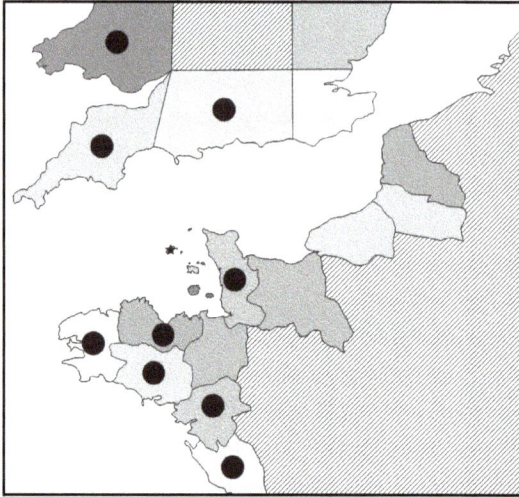

This anemone is wide-spread throughout NW Europe where it is commonly associated with harbours and ports, especially those in lower salinity settings such as estuaries. Although recorded from marinas in St Peter Port, it has not yet been discovered elsewhere in the Channel Islands. It has been reported as a fouling organism in some ports but it is unlikely to present any problems within the Channel Islands.

# Mnemiopsis leidyi
## Sea Walnut - ctenophora

Threat score: 60

| ES | Hab | Tox | Econ |
|----|-----|-----|------|
| 3 | 4 | 1 | 5 |

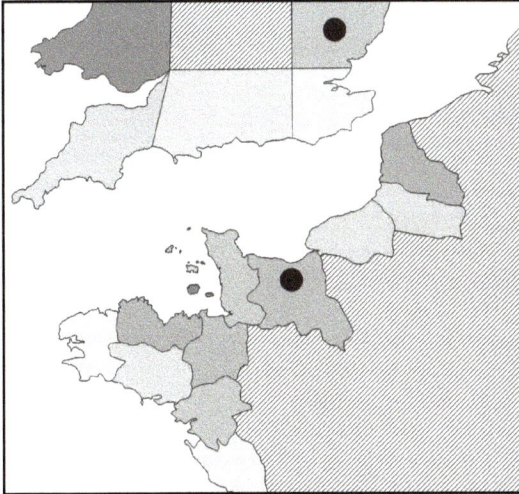

Native to North and South America, *M. leidyi* was introduced into the Black Sea in the 1980s probably via ships' ballast. In 2005 specimens were found in Denmark and by 2016 it had been recorded in England and just east of Cherbourg. *M. leidyi* is a voracious predator of zooplankton (including of fish eggs and larvae) which can outcompete native species to the extent that it has already had an impact on the pelagic ecology of the Black Sea. It can tolerate a wide range of salinities and temperatures and may be more of an issue for estuaries and enclosed environments than open coasts. The ability of *M. leidyi* to predate on eggs and larvae led to economic and ecological impacts within the Black Sea and the species is considered to be a high threat to north-west Europe by Roy *et al.* (2014). Recent survey work suggests that it is now common in the eastern English Channel and it is probable that it will be found in Channel Island waters in the near future.

# *Pseudodactylogyrus anguillae*
## Platyhelminthes

Threat score: 64

| ES | Hab | Tox | Econ |
|----|-----|-----|------|
| 2 | 4 | 4 | 2 |

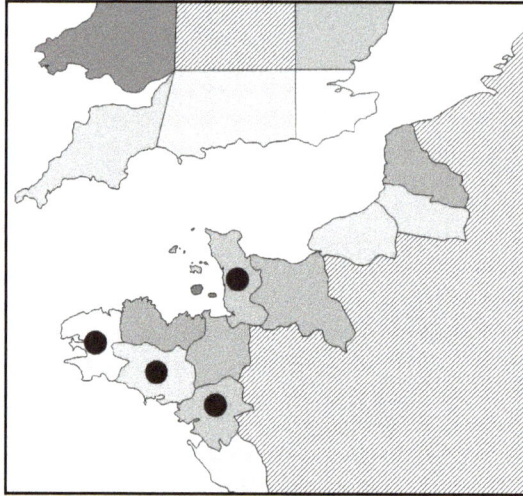

A parasitic flatworm originally from the North-west Pacific but which was discovered in 1977 on commercial eel farms in the western Soviet Union. During the 1980s *P. anguillae* has spread to other eel aquaculture sites in Europe including in Brittany and Denmark. The parasite causes gill infections in the European Eel (*Anguilla anguilla*) and can have a serious effect on commercial production. The European Eel is rare on Jersey and the species is not subject to aquaculture within the Channel Islands. The likelihood of *P. anguillae* establishing itself locally is remote.

# *Koinostylochus ostreophagus*
## Platyhelminthes

Threat score: 24

| ES | Hab | Tox | Econ |
|----|-----|-----|------|
| 2 | 2 | 3 | 2 |

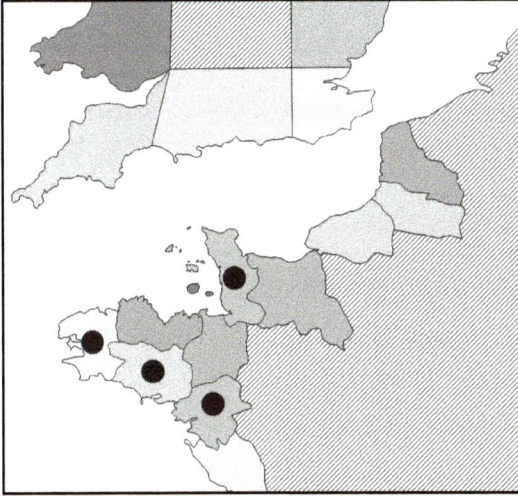

Native to the North-west Pacific, this polyclad flatworm is a predatory parasite on juvenile oysters and mussels. *K. ostreophagus* has become widespread across the globe and its local occurrence would appear to be synonymous with aquaculture, especially oysters. Although frequently recorded and sometimes said to reach 'pest proportions', the effects of *K. ostreophagus* on commercial oyster production are not entirely clear.

The first European specimens were found during the early 1970s and it has been widely reported since including from most of the oyster production areas on the Normandy and Brittany coasts. Given its widespread occurrence and association with aquaculture, it seems probable that *K. ostreophagus* will be present within the Channel Islands but it does not appear to represent a serious threat to the local ecology or oyster farming industry.

# *Anguillicola crassus*
## Nematode

Threat score: 15

| ES | Hab | Tox | Econ |
|----|-----|-----|------|
| 1 | 5 | 3 | 1 |

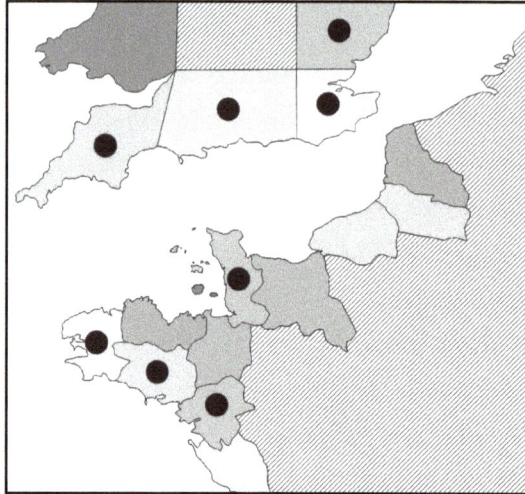

A parasitic nematode of the European Eel (*Anguilla anguilla*) that was introduced to Europe in 1987 from South-east Asia. It is usually associated with fresh and brackish waters but is occasionally found in marine specimens. This parasite is listed by some people as being a major menace to the health of the European Eel population due the amount of blood that it can adsorb. It has even been suggested as a possible cause for the decline in eel stocks. *A. crassus* is found along the Atlantic French coast and is probably present in Jersey's European Eel population but it is of low threat to human health.

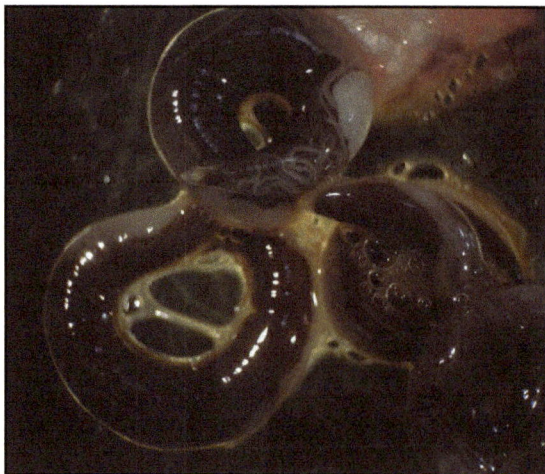

# *Boccardia semibranchiata*
## Annelida

Threat score: 2

| ES | Hab | Tox | Econ |
|----|-----|-----|------|
| 2  | 1   | 1   | 1    |

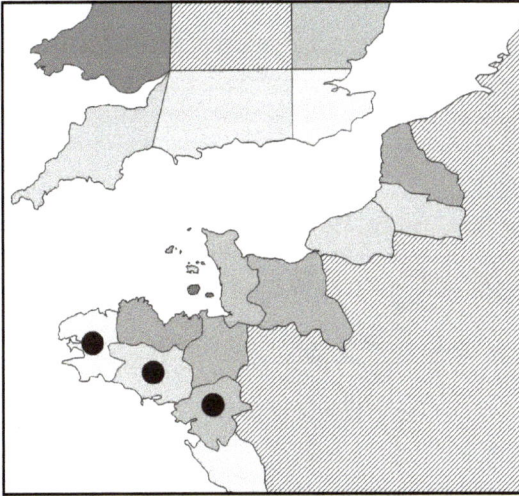

A mud worm that is native to the Mediterranean but which was identified on oyster farms in Normandy in 1990 and Southern Brittany in 1999. The species is not regarded as a threat and seems primarily to be transported with oyster stock. Biosecurity measures within the aquaculture industry may lessen (but not exclude) the possibility of its arrival in the Channel Islands.

# *Hydroides* spp.
## Annelida

Threat score: 4

| ES | Hab | Tox | Econ |
|----|-----|-----|------|
| 2 | 2 | 1 | 1 |

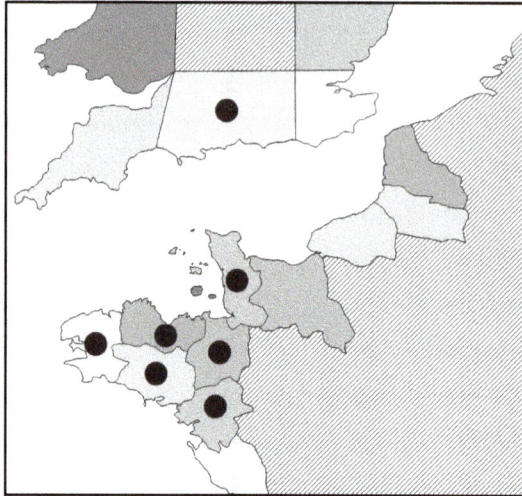

This page covers three species of calcareous tube worm originally from the SW Pacific (*H. elegans*) NW Pacific (*H. ezoensis*) and the eastern USA (*H. dianthus*). Of the three, *H. dianthus* was the earliest to arrive in Europe having been reported from Ile de Ré (Biscay) in 1927 but has since spread across Brittany, Normandy and other coasts including the UK. *H. elegans* was first reported in 1937 and *H. eozensis* in 1976 but while both species have records from the southern coast of England neither has been identified in the Normano-Breton Gulf.

These species have a wide environmental tolerance and will compete with local tube worms but they are not noted as a serious threat. As these are already regionally widespread, one or more of these tubeworms may already be present in the Channel Islands.

# *Goniadella gracilis*
## Annelida

Threat score: 4

| ES | Hab | Tox | Econ |
|----|-----|-----|------|
| 2 | 2 | 1 | 1 |

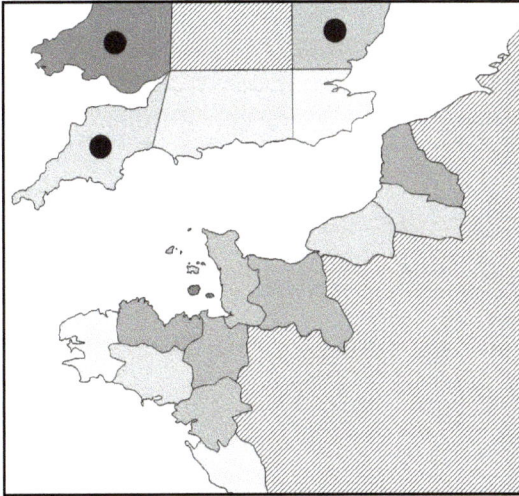

First reported from Liverpool in 1970, *G. gracilis* has since spread to many parts of the Irish Sea. Although still some distance from the Channel Islands, it is possible that it may reach the English Channel in coming decades. It is not thought to be a serious threat to the Channel Islands.

# *Desdemona ornata*
## Annelida

Threat score: 12

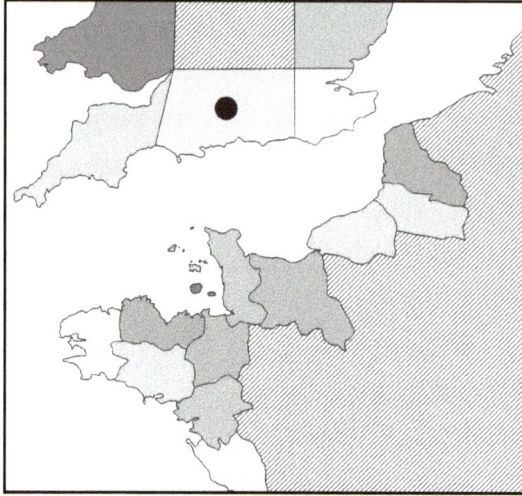

When first identified at Southampton in 1997 *D. ornata* was already present in densities of up to 12,000 m$^{-2}$. A single specimen was afterwards found in south Devon. Current records suggest that *D. ornata* may have a preference for organically enriched muddy littoral sites and in Southampton highest densities have been associated with sewage outfalls. It may prefer lower estuary habitats of a sort that do not occur in the Channel Islands but its presence in some muddier harbours cannot be excluded. Although not regarded as an environmental or economic threat at present, the densities recorded at Southampton suggest that it has the potential to affect ecosystems.

# *Ficopomatus enigmaticus*
## Annelida

Threat score: 4

| ES | Hab | Tox | Econ |
|----|-----|-----|------|
| 1  | 2   | 1   | 2    |

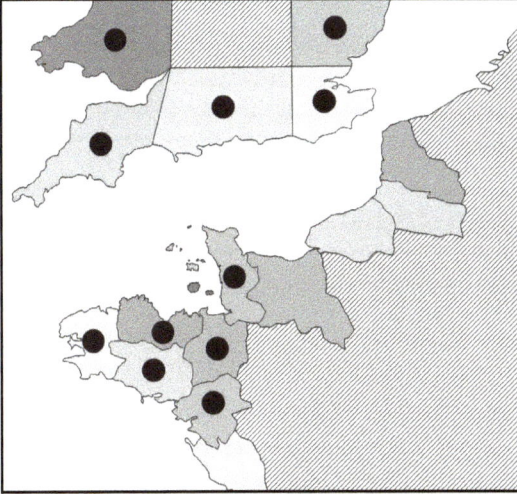

An Australian calcareous tube worm that was first recorded in London's docklands in 1922. It has been recorded at most major ports in Normandy, Brittany and the UK where it is sometimes listed as a fouling organism. *F. enigmaticus* has a preference for estuarine and brackish waters and attaches itself to hard substrates. It has not been reported from the Channel Islands and it may be that a lack of lower salinity environments will prohibit its establishment.

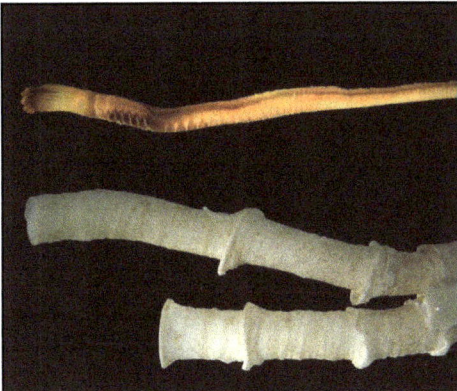

# *Neodexiospira brasiliensis*
## Annelida

Threat score: 4

| ES | Hab | Tox | Econ |
|----|-----|-----|------|
| 2 | 2 | 1 | 1 |

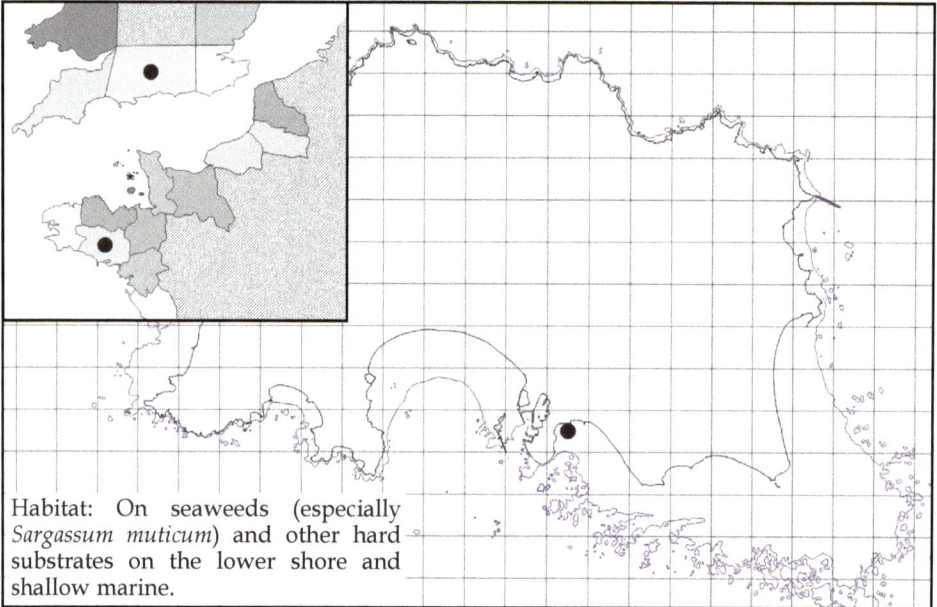

Habitat: On seaweeds (especially *Sargassum muticum*) and other hard substrates on the lower shore and shallow marine.

First reported in Portsmouth Harbour in 1974, it is thought that this small species of tube worm may have been spread by being attached to specimens of the non-native brown seaweed *Sargassum muticum*.

First identified in Jersey at La Collette in 1987 by visiting lecturers from Portsmouth University, it has not since been identified on the island and has no records at all from anywhere else in the region. Its present status in the Channel Islands is unknown but it is not thought to present a significant threat to the local environment. Other European records are sporadic and rare.

# *Pileolaria berkeleyana*
## Annelida

Threat score: 4

| ES | Hab | Tox | Econ |
|----|-----|-----|------|
| 2  | 2   | 1   | 1    |

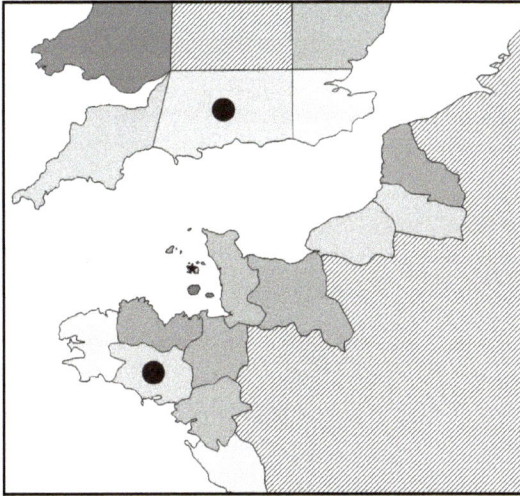

A small polychaete worm from the Pacific Ocean that was found in Portsmouth Harbour in 1974, in southern Brittany in 1982 and from a single specimen on *Sargassum muticum* in St Helier, Jersey, in 2007. The rate of spread is slow and *P. berkeleyana* is not considered to be a threat to native habitats or species. Its present status in the Channel Islands is unknown.

# *Ammothea hilgendorfi*
## Sea Spider - chelicerata

Threat score: 16

| ES | Hab | Tox | Econ |
|----|-----|-----|------|
| 4 | 4 | 1 | 1 |

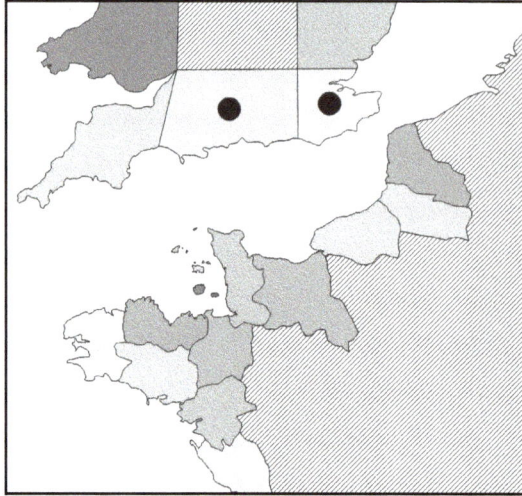

Native to the North-west Pacific, this species of sea spider was first recorded in Southampton waters in 1978 and then shortly afterwards in Venice. In 2010 *A. hilgendorfi* was found in Poole Harbour from where it has spread to the neighbouring seashore to become 'super-abundant'. It has since been recorded from several North Sea locations including the Essex coast and the Netherlands. Given that many southern England non-native species have found their way to the Channel Islands, the establishment of *A. hilgendorfi* locally is a possibility. Its effect on habitats and local species is being assessed but this species should be on a watch list.

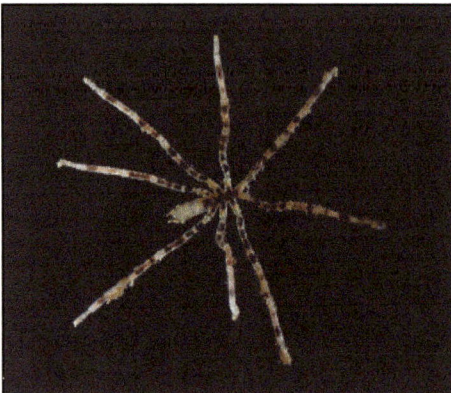

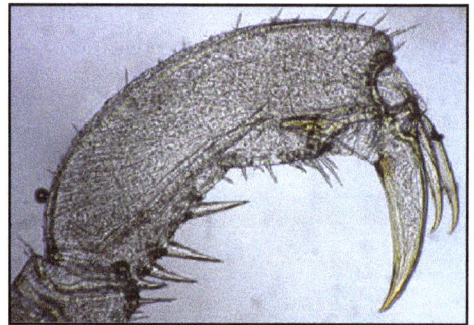

*Left: Whole animal. Right: Distal part of right hind leg with claws and auxillary claws.*

# *Acartia tonsa*
## Copepod

Threat score: 4

| ES | Hab | Tox | Econ |
|----|-----|-----|------|
| 2  | 2   | 1   | 1    |

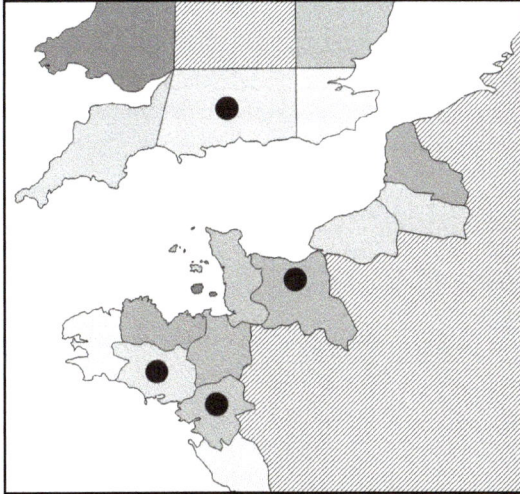

A copepod which can tolerate a wide range of temperatures and salinities but which is particularly common in estuary situations. It is geographically widespread which may be due to it habing been transported in ballast water. Although *A. tonosa* can survive in coastal waters, it is most abundant in lower salinity situations but even in these conditions it is not listed as being a threat. This species is farmed commercially in Guernsey but has not been reported from the marine environment around any of the islands. It is not listed as a potential threat and is unlikely to pose any problems.

# *Mytilicola intestinalis*
## Copepod

Threat score: 16

| ES | Hab | Tox | Econ |
|----|-----|-----|------|
| 2 | 2 | 1 | 4 |

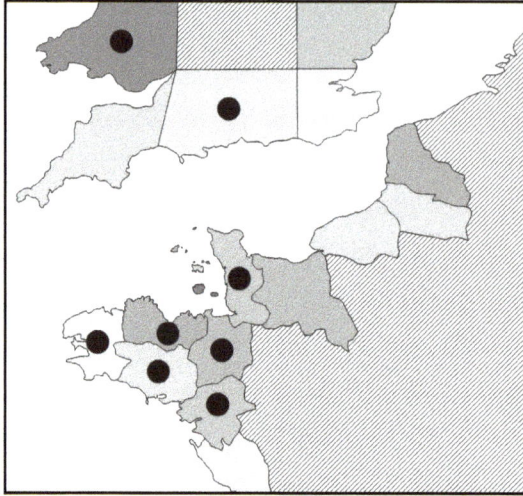

A parasitic copepod from the Mediterranean that is associated with mussels. The first northern European record was in 1937 in the UK but by the 1950s it had spread (probably via aquaculture) to the Atlantic French coast. *M. intestinalis* is a noted pest to mussel aquaculture and has been recorded from all Brittany and Normandy coasts. Jersey has a small mussel aquaculture industry on the east coast but this does provide a potential habitat for this species although it has never actually been recorded.

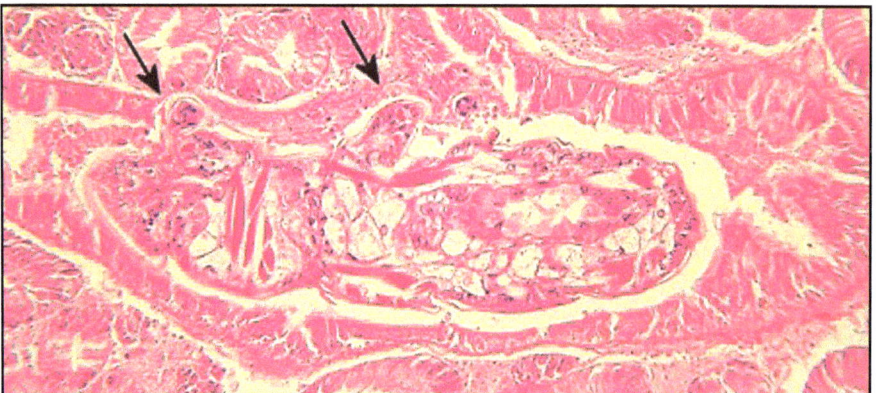

*A Mytilicola copepod inside the intestinal tract of a bivalve mollusc.*

# *Mytilicola orientalis*
## Red Worm - copepod

Threat score: 24

| ES | Hab | Tox | Econ |
|----|-----|-----|------|
| 2  | 2   | 3   | 2    |

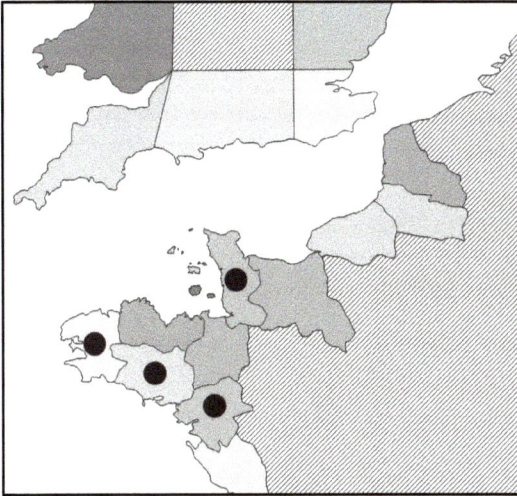

A parasitic copepod of oysters from the North-west Pacific that was introduced into Brittany in 1977 via aquaculture. It has since spread widely and its presence is suspected in many areas associated with oyster aquaculture including in the UK.

Its principal host species are *Crassostrea edulis* and *Mytilus edulis* and although *M. orientalis* is not regarded as a threat to health or the environment caution is recommended. Given its association with aquaculture in Brittany and Normandy, it may be present in Jersey's oyster production areas too. Listed as a medium threat by Roy *et al.* (2014).

# *Pseudomyicola spinosus*
## Copepod

Threat score: 24

| ES | Hab | Tox | Econ |
|----|-----|-----|------|
| 2 | 2 | 3 | 2 |

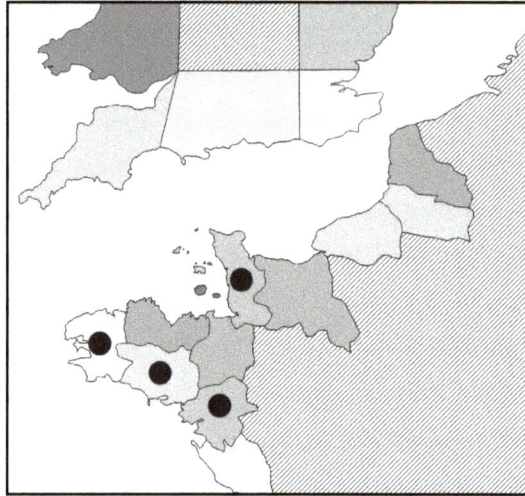

A parasitic copepod from the Pacific Ocean associated with mussels. First reported in Normandy in 1963, it has since been found on several locations along the French coast. Its movement is closely associated with aquaculture and it is not considered to be a threat to the wider environment. Its status in the Channel Islands is not known.

# Myicola ostreae
## Copepod

Threat score: 16

| ES | Hab | Tox | Econ |
|----|-----|-----|------|
| 2 | 2 | 2 | 2 |

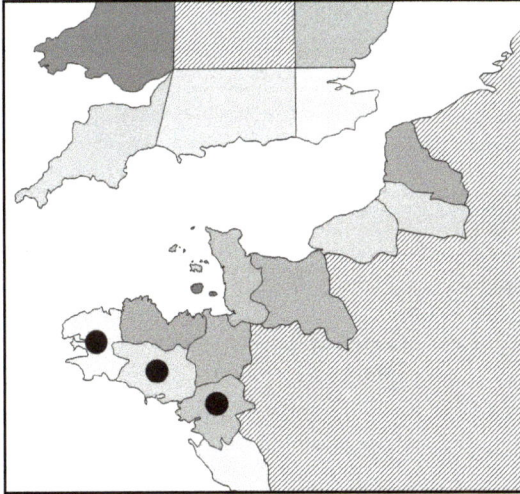

A parasitic copepod native to the North-west Pacific but which was introduced into southern Brittany in the early 1970s via aquaculture. It has since been recorded from a number of other locations along the Atlantic coast, normally in association with oyster farms. *M. ostreae* parasitizes the gills of bivalve molluscs, especially *Crassostrea gigas*, and its movement within Europe seems to be directly linked to the translocation of oysters. It has not been recorded from the Channel Islands but given the legacy of aquaculture, it may conceivably be present although it is not thought to be an environmental or economic threat. The probable movement of this species with its host suggests that biosecurity represents the best means of preventing its spread.

# *Hesperibalanus fallax*
## Barnacle

Threat score: 8

| ES | Hab | Tox | Econ |
|----|-----|-----|------|
| 2 | 2 | 1 | 2 |

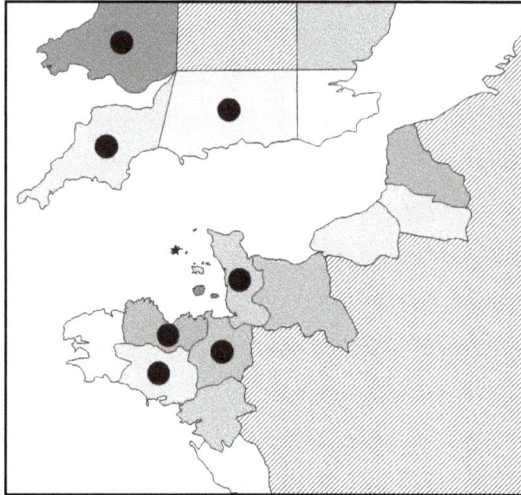

A specialist barnacle that generally attaches itself to the shells of whelks (*Buccinum undatum*) and Queen Scallops (*Aequipecten opercularis*). Native to North-west Europe but discovered in southern England in 1994 and southern Brittany in 2000; it is now thought to be widespread in many parts of the English Channel. In 2001 a string of used lobster pots bought from Guernsey and imported into Holland were found to have a number of *H. fallax* specimens growing on them (Southward *et al.*, 2004).

Although regarded as a potential fouling species, *H. fallax* is unlikely to present a threat to local species. Assessments in the UK have focused on the examination of lobster and crab pots which seem to provide a favourable substrate for *H. fallax*. Similar assessments could be carried out in the Channel Islands to establish its presence and distribution.

# *Austrominius modestus*
## Australian Barnacle

Threat score: 4

| ES | Hab | Tox | Econ |
|----|-----|-----|------|
| 2 | 2 | 1 | 1 |

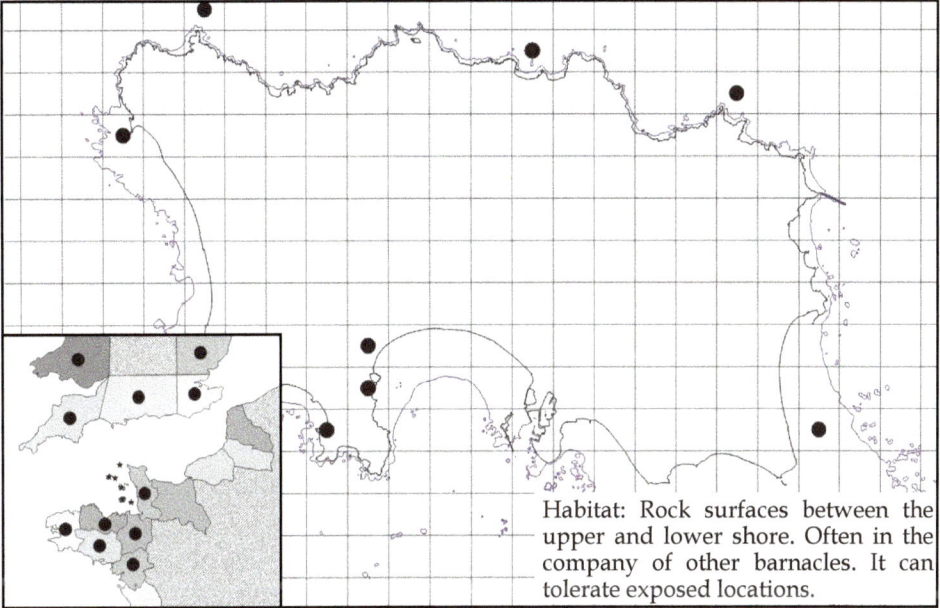

Habitat: Rock surfaces between the upper and lower shore. Often in the company of other barnacles. It can tolerate exposed locations.

A non-native species that was first identified from Chichester Harbour (UK) in 1945. It spread rapidly to many parts of Europe and had reached Guernsey, and probably the other Channel Islands, by 1958 although the first Jersey record was not until 1977 at Les Minquiers.

The Australian Barnacle is not common on Channel island seashores and tends to be found as individual specimens, especially on the middle and upper lower shore. It is not considered to be a threat to the local environment.

Left: *Specimens of* A. modestus *under a rock at L'Étacq, Jersey.* Right: *Specimens on the seashore in France.*

# *Amphibalanus amphitrite*
## Purple Acorn Barnacle

Threat score: 6

| ES | Hab | Tox | Econ |
|----|-----|-----|------|
| 2  | 3   | 1   | 1    |

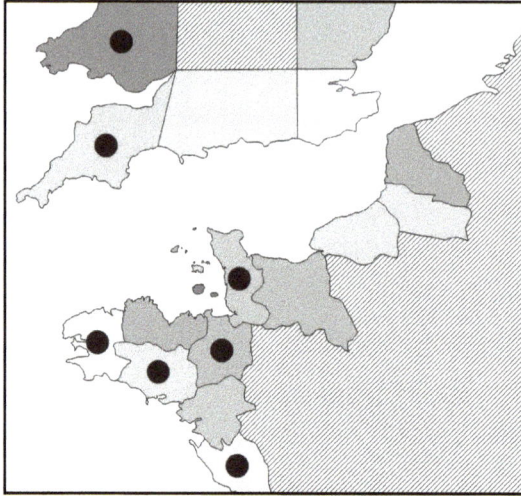

A native barnacle from the South-west Pacific, *A. amphitrite* was present in Europe from at least the 1920s when it was recorded from several major ports. Noted as a potential competitive species, *A. amphitrite* has spread to most temperate parts of the world where it has colonised a variety of marine and lower salinity habitats. Given its widespread recording on adjacent French coasts, the presence of this species in Channel Island waters is a distinct possibility.

# *Amphibalanus eburneus*
## Ivory Barnacle

Threat score: 4

| ES | Hab | Tox | Econ |
|----|-----|-----|------|
| 2  | 2   | 1   | 1    |

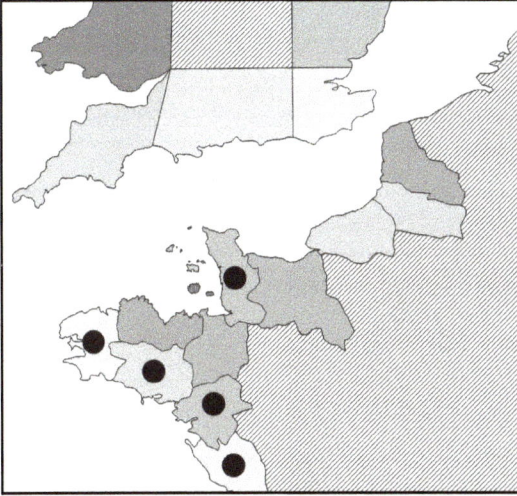

A medium-sized barnacle from the eastern USA, *B. eburneus* was first recorded in southern Brittany in the 1940s followed by sporadic reports from individual locations along other parts of the Atlantic coast. It is not considered to be a threat in temperate waters and may struggle to establish itself outside of the tropics. It is a species that could become more prevalent in the English Channel as global sea temperatures rise and so may eventually be found in the Channel Islands. However, its similarity to native species could make it difficult to identify by non-specialists.

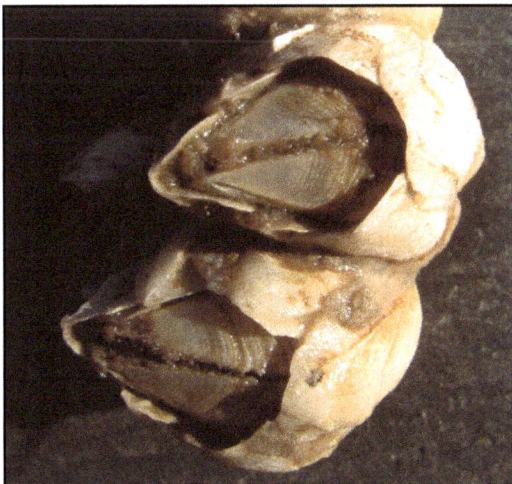

# *Amphibalanus improvisus*
## Barnacle

Threat score: 4

| ES | Hab | Tox | Econ |
|----|-----|-----|------|
| 2  | 2   | 1   | 1    |

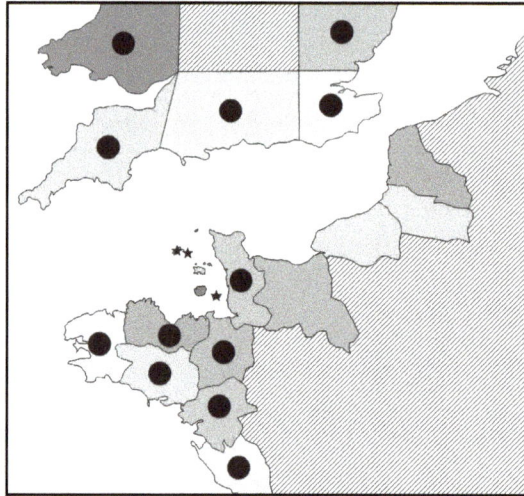

A North American species that was established in London's docklands by the 1850s and which has spread widely since. There is an early record from Sark (1865) and from Brittany in the 1870s. This is a species that is principally (but not exclusively) associated with lower salinities (below 20 ppt) and it is a possibility that the lone Sark record is a misidentification. Although noted as a fouling species, *B. improvisus* is unlikely to become established locally but the recovery of specimens from buoys off Belgium in the 1990s suggests that it might be present in local waters.

# *Amphibalanus reticulatus/A. variegatus*
## Reticulated Barnacles

Threat score: 8

| ES | Hab | Tox | Econ |
|----|-----|-----|------|
| 2 | 2 | 1 | 2 |

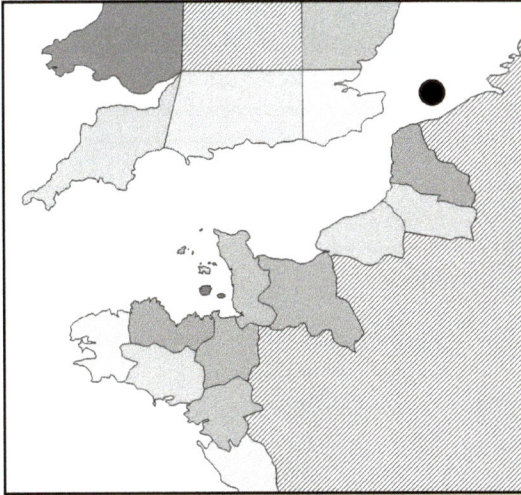

Two small species of barnacle native to the North-west Pacific (*A. reticulatus*) and South Pacific (*A. variegatus*). Specimens were reported from buoys moored of Belgium in 1997 but there do not seem to be other records from the English Channel. Their occurrence in the Channel Islands is possible but at present these are not considered to be a general threat to local ecology or species.

*Left: Amphibalanus reticulatus*
*Right: Amphibalanus variegatus*

# *Balanus trigonus*
## Triangular Barnacle

Threat score: 8

| ES | Hab | Tox | Econ |
|----|-----|-----|------|
| 2  | 2   | 1   | 2    |

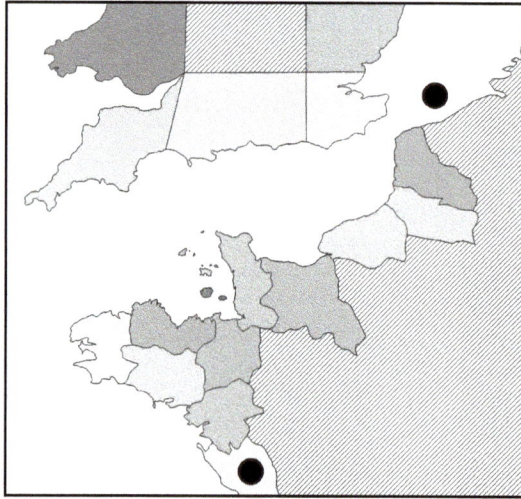

Native to the Indo-Pacific area, this medium-sized barnacle has spread widely and is considered to be cosmopolitan across tropical and temperate seas. It has been reported from many European locations including the English Channel where specimens were recovered from buoys off Belgium in 1997 (see also *Amphibalanus reticulatus* and *A. variegatus*). Widely regarded as a fouling species in some parts of the world the threat presented by this species in the English Channel seems to be low. Not reported from the Channel Islands but its occurrence in local waters should be considered a possibility.

# *Megabalanus coccopoma*
## Titan Acorn Barnacle

Threat score: 27

| ES | Hab | Tox | Econ |
|----|-----|-----|------|
| 3  | 3   | 1   | 3    |

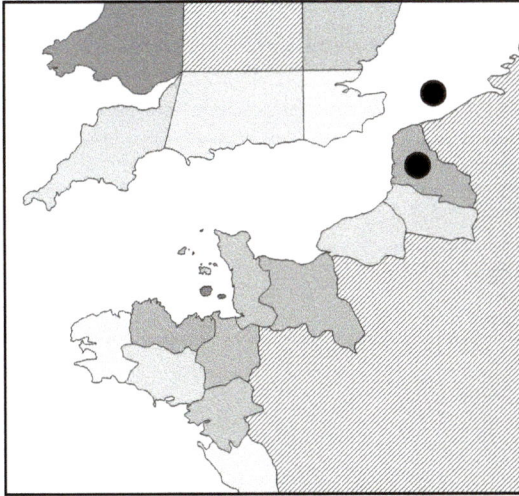

Native to the Pacific coasts of South and Central America, *M. coccopoma* has recently established itself at several locations worldwide. It was reported from buoys anchored off Belgium in 1997 and has since been reported from ports in northern France although it does not yet seem to be in the Normano-Breton Gulf. This species is large, is noted as a fouling pest and appears to be being transported via shipping and floating objects. It is listed as a medium threat to the British Isles by Roy *et al.* (2014) and is a species whose spread should probably be monitored. Its size (up to 5 cm in height and width) should make it easy to spot.

# *Megabalanus tintinnabulum*
## Sea Tulip

Threat score: 27

| ES | Hab | Tox | Econ |
|----|-----|-----|------|
| 3  | 3   | 1   | 3    |

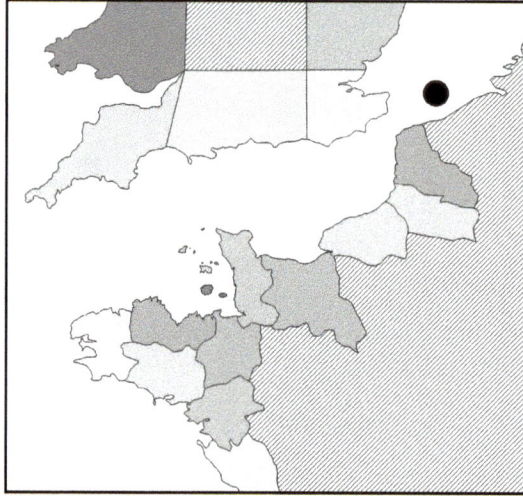

Native to the Indo-Pacific and possibly West Africa, *M. tintinnabulum* has spread widely through the tropics and sub-tropics and was first identified in Europe in 1997 from the same location off Belgium as *M. coccopoma*. The description given for *M. coccopoma* applied to this species also. Listed as medium threat to Britain in Roy *et al.* (2014).

# *Eusarsiella zostericola*
## Ostracod

Threat score: 8

| ES | Hab | Tox | Econ |
|----|-----|-----|------|
| 2 | 2 | 1 | 2 |

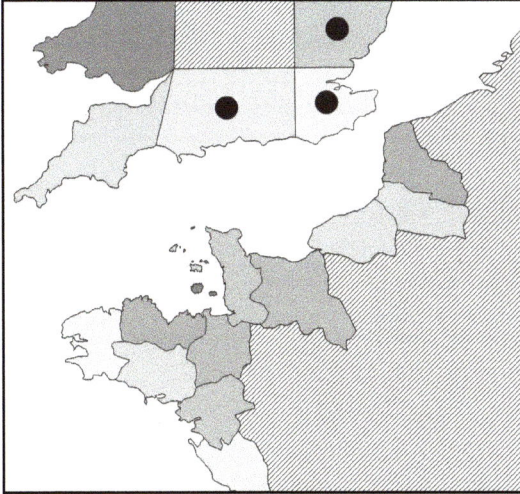

An ostracod from the eastern USA associated with the commercial culture of the oyster *Crassostrea gigas*. Originally imported into the UK in the 1870s, this species was thought to be slow dispersing and until recently only two established colonies were known. However, a population of *E. zostericola* was found offshore in 2013 in the Netherlands suggesting that it is spreading. Its minute size and the need for specialist identification means that this species could have been overlooked and is further distributed than records would suggest. It is not considered to be a serious threat but could eventually be found in the Normano-Breton Gulf.

# *Odontodactylus scyllarus*
## Peacock Mantis Shrimp

Threat score: 4

| ES | Hab | Tox | Econ |
|----|-----|-----|------|
| 2  | 2   | 1   | 1    |

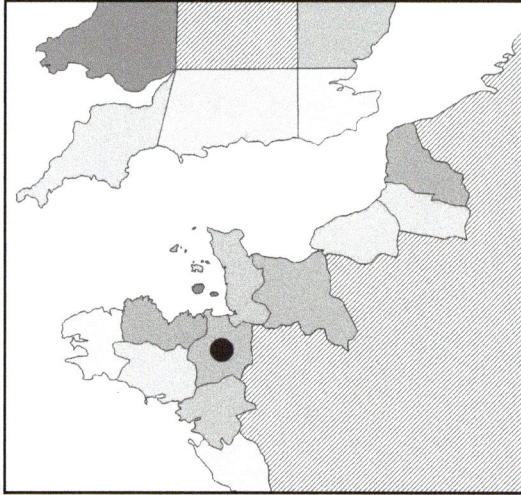

A mantis shrimp native to the Indo-Pacific Ocean. A single specimen was photographed by a diver off St Malo in 2009 but it does not seem to have been recorded since. It is thought that the specimen may have escaped from an aquarium or have been deliberately released. It is unlikely that this tropical species could establish itself in the English Channel.

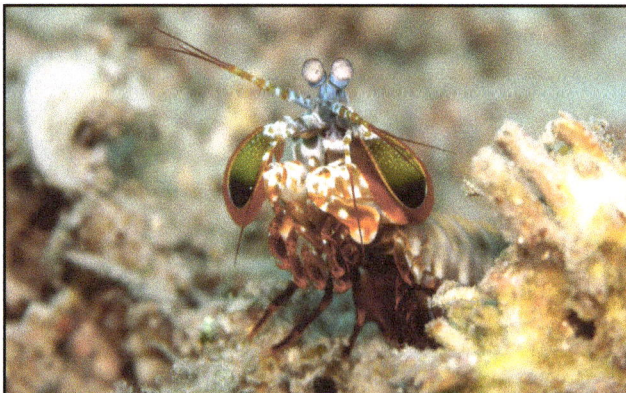

# *Monocorophium sextonae*
## Amphipod

Threat score: 4

| ES | Hab | Tox | Econ |
|----|-----|-----|------|
| 2 | 2 | 1 | 1 |

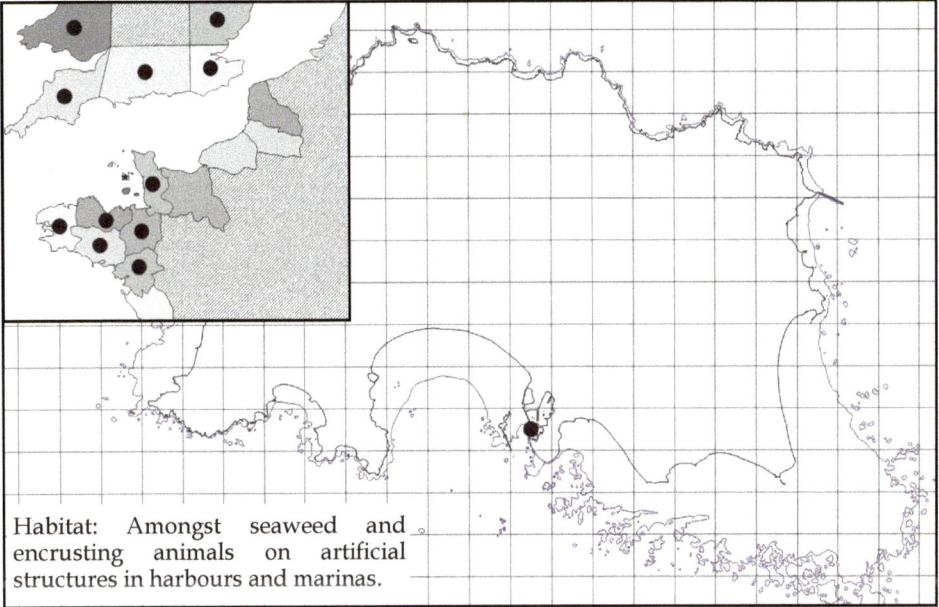

Habitat: Amongst seaweed and encrusting animals on artificial structures in harbours and marinas.

Widespread in the English Channel and other regions since the 1930s and first reported from the Channel Islands in the 1950s. *M. sextonae* is common in St Helier Marina where it builds tubes among the weeds and other organisms encrusting the pontoons. This species is not regarded as a serious threat to local habitats or species. Its small size makes it difficult to identify without a microscope.

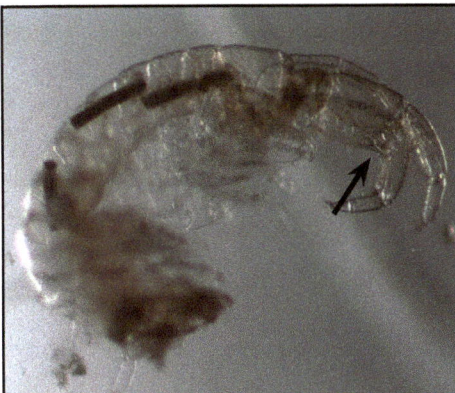

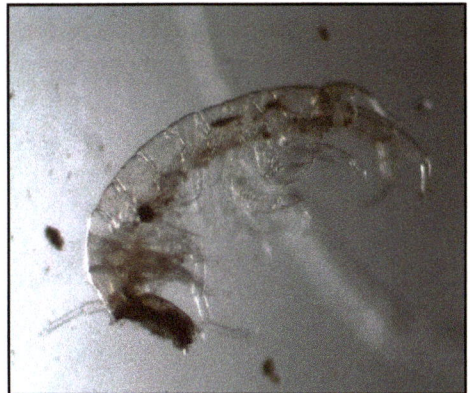

*Left: Male specimen from St Helier Marina. Toothed antenna is arrowed.*
*Right: Female specimen from St Helier Marina.*

# *Grandidierella japonica*
## Amphipod

Threat score: 12

| ES | Hab | Tox | Econ |
|----|-----|-----|------|
| 3 | 4 | 1 | 1 |

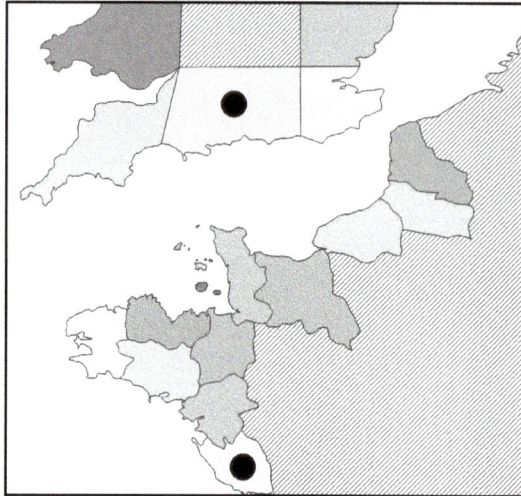

When first identified at Southampton in 1997 *G. japonica* was already present in densities of up to 5,800 m$^{-2}$. This species may prefer lower estuary habitats of a sort that do not occur in the Channel Islands but its presence in some of our muddier harbours cannot be excluded. Although not regarded as an environmental or economic threat at present, it probably outcompetes the local species *Aora gracilis* and the densities recorded at Southampton suggest that it has the potential to affect ecosystems.

# *Caprella mutica*
## Japanese Skeleton Shrimp

Threat score: 6

| ES | Hab | Tox | Econ |
|----|-----|-----|------|
| 2 | 3 | 1 | 1 |

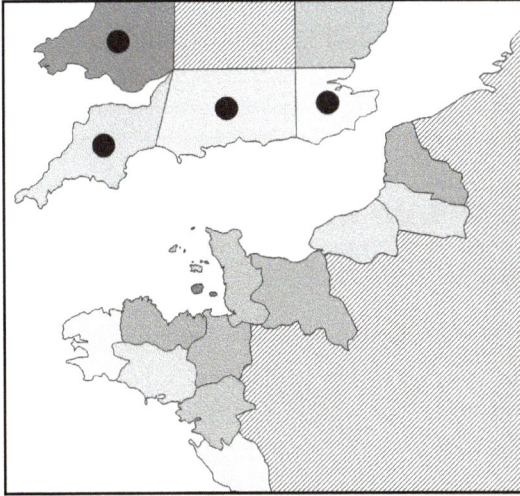

Native to the North-west Pacific, *C. mutica* was recorded in Scotland in 2000 but has since been discovered at several locations in the North Sea, Irish Sea and English Channel although not, as yet, in Normandy and Brittany. It is associated with centres of human activity such as marinas and harbours where it will live amongst seaweeds, hydroids, etc. It was recently discovered at several southern England ports which leaves open the possibility that it may be carried to the Channel Islands although deliberate searches in Jersey's marinas have not yielded any specimens as yet. *C. mutica* is not considered to be a threat.

# *Limnoria quadripunctata*
## Four-spotted Gribble

Threat score: 12

| ES | Hab | Tox | Econ |
|----|-----|-----|------|
| 2  | 2   | 1   | 3    |

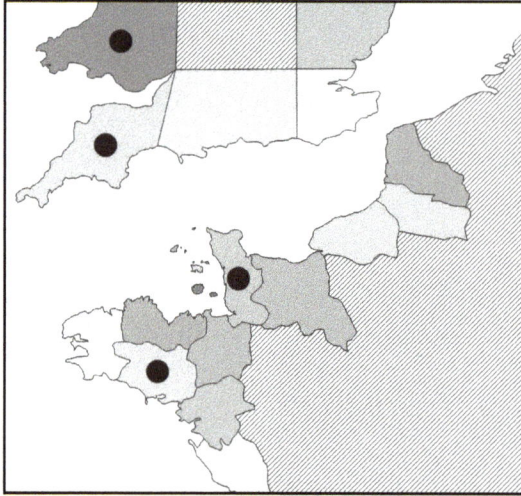

Native to the southern Pacific ocean, this species has been present in European water since at least 1930 and is now widespread in temperate waters across the globe. It is a detritivore that can do serious damage to wooden piers, hulls, etc. There are historical reports from Granville and Chausey but no actual Channel Island records. It could exist within the islands but, as with many boring marine species, a general lack of wooden structures limits opportunities for establishment.

# *Limnoria tripunctata*
## Three-spotted Gribble

Threat score: 12

| ES | Hab | Tox | Econ |
|----|-----|-----|------|
| 2  | 2   | 1   | 3    |

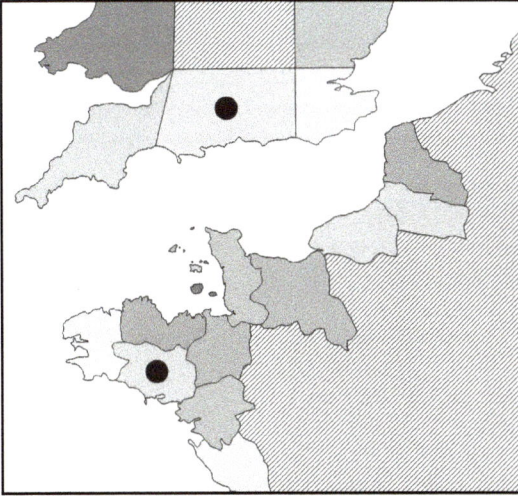

Native to the Pacific Ocean, *L. tripunctata* is now widespread across the globe. It was first reported in the English Channel in the 1950s and identified in southern Brittany in 2009. Although widespread, it is not often sought and its local distribution is poorly understood. It may well be wider distributed than records suggest and might be present in Channel Islands waters although a lack of available wooden media into which it can bore may limit its presence.

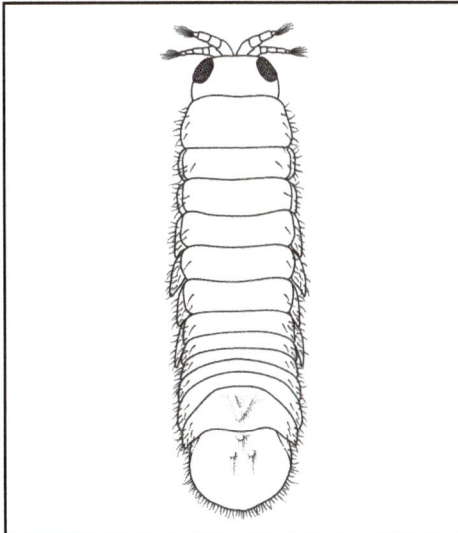

# *Penaeus japonicus*
## Kuruma Prawn

Threat score: 8

| ES | Hab | Tox | Econ |
|----|-----|-----|------|
| 2  | 4   | 1   | 1    |

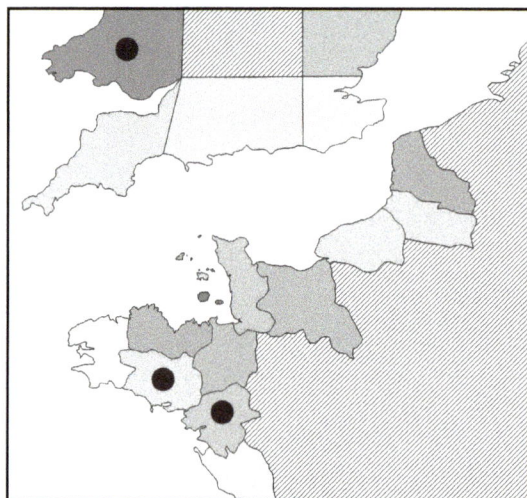

Native to the Indo-Pacific Ocean area, this large (up to 24 cm) species of prawn has been established in the eastern Mediterranean for some decades but has reports from southern Brittany in 1980 and 2005. There have also been isolated reports from other Atlantic locations including several from the western English Channel. However, this species appears to require warm waters (at least 24°C) in order to breed and it is not thought to be established in northern Europe. In areas where it has become established, *P. japonicus* has out-competed local prawn species and predicted rises in local sea temperatures may make this species a longer-term threat to the European Atlantic coast. *P. japonicus* is a distinctive burrowing species of the extreme lower shore and shallow marine; it is largely nocturnal and, given other records in the English Channel, it is possible that specimens may be reported from the Channel Islands although a lack of commercial fishing for prawns lessen the likelihood of specimens being captured.

# *Palaemon macrodactylus*
## Oriental Prawn

Threat score: 6

| ES | Hab | Tox | Econ |
|----|-----|-----|------|
| 2 | 3 | 1 | 1 |

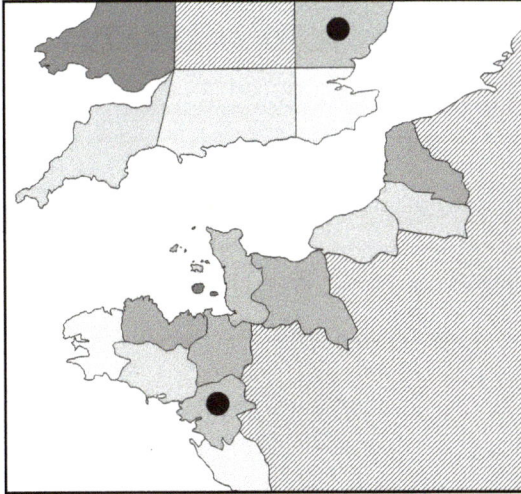

A large prawn from North-west Europe that was found in the UK in 1992 and southern Brittany in 2006. It has the potential to spread further and, although it can survive in fully marine conditions, it is most abundant in estuaries and other lower salinity habitats. *P. macrodactylus* may compete with native species but as an essentially estuarine species, it is unlikely to become widely established in the Channel Islands.

# *Homarus americanus*
## Canadian Lobster

Threat score: 30

| ES | Hab | Tox | Econ |
|----|-----|-----|------|
| 2 | 5 | 1 | 3 |

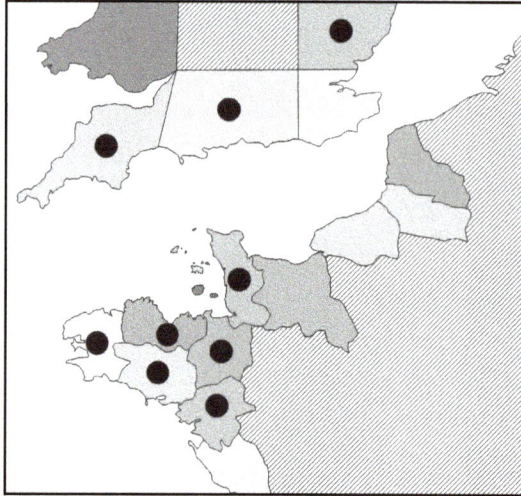

Similar to the European lobster but larger and heavier, *H. americanus* is native to the east coast of the USA but has been recorded in the English Channel since at least 1988. Specimens were identified off Normandy (Manche) in 2005 and others have since been found off northern and southern Brittany coasts. *H. americanus* is very similar to the European lobster (*H. gammarus*) and it is presumed that many specimens have been caught and not recorded.

There have been no confirmed reports of *H. americanus* in Channel Island waters but given its presence elsewhere in the Normano-Breton Gulf it is probable that the species occurs locally. It is listed a high threat to the British Isles by Roy *et al.* (2014) with the principal concerns being a potential to outcompete the European lobster (and possibly the Brown Crab, *Cancer pagurus*) and a proven ability to hybridise with native lobsters. Aside from possible environmental impacts, there could also be impacts on the commercial shellfish industry. Assessment of *H. americanus* locally could be achieved by making fishermen aware of the species and how to identify it or via targeted research during annual lobster trials.

# *Asthenognathus atlanticus*
## Crab

| ES | Hab | Tox | Econ |
|----|-----|-----|------|
| 2  | 2   | 1   | 2    |

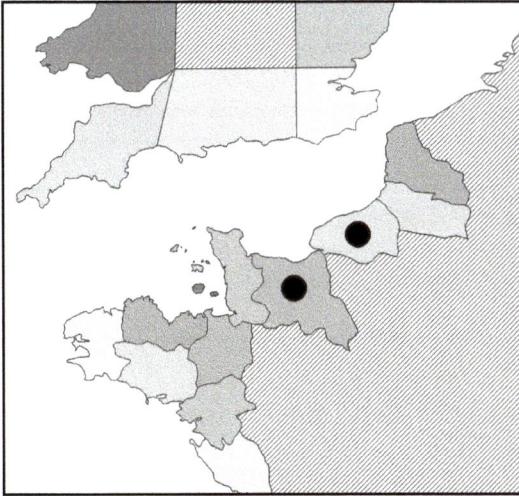

The native range of *A. atlanticus* is believed to stretch from the West African Atlantic coast to the northen part of the Bay of Biscay, where it is considered to be rare. In 1921 specimens were reported at Roscoff, north Brittany and, since 2000, it has been found at St Malo, Cornwall and the eastern part of the Bay of Seine, Normandy. This patchy distribution can be typical of a non-native species and suggests that *A. atlanticus* is being moved by human agencies rather than naturally extending its range.

*A. atlanticus* is not yet known from the Channel Islands and its preference for offshore muddy sediments (which are not common in the islands) may restrict its occurrence. However, the St Malo specimens were found in seagrass (*Zostera* spp.) habitats and so its establishment locally is a possibility.

# *Hemigrapsus sanguineus*
## Asian Shore Crab

Threat score: 30

| ES | Hab | Tox | Econ |
|----|-----|-----|------|
| 3 | 5 | 1 | 2 |

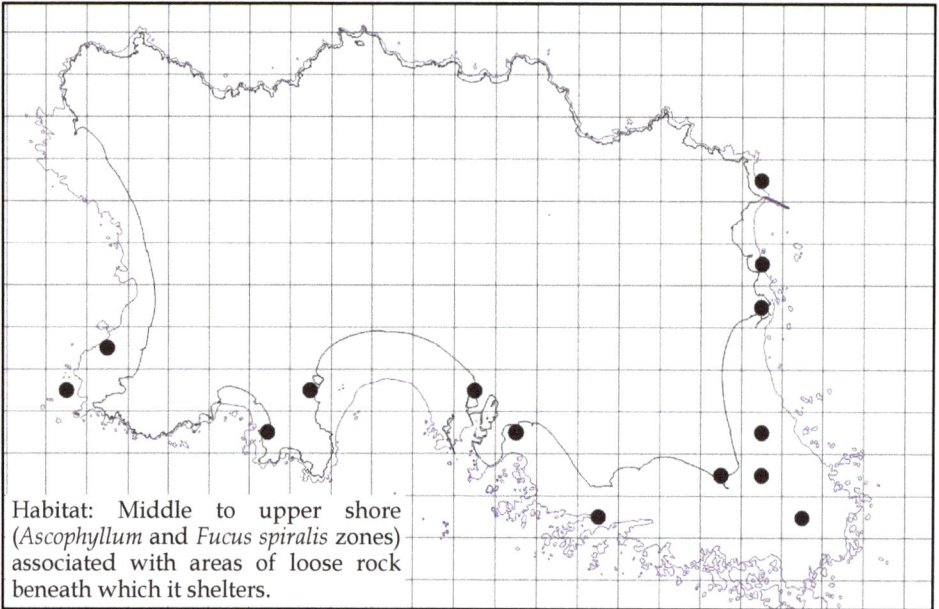

Habitat: Middle to upper shore (*Ascophyllum* and *Fucus spiralis* zones) associated with areas of loose rock beneath which it shelters.

*Hemigrapsus sanguineus* was first reported in 1998 from the Netherlands from where it has spread westwards. By 1999 had been found at Le Havre and it would appear to have then worked its way southwards being reported from Guernsey and then Jersey in the same week in May 2009. It was not reported from the UK until 2014 and Brittany until July 2016 when a single specimen was recovered at Pors-Even en Ploubazlanec on the Côtes d'Armor.

Following its discovery in Jersey in 2009 *H. sanguineus* was rarely reported on the seashore and, despite deliberate searching, could only be reliably found adjacent to the power station water outlet at La Collette. (NB: The first UK specimen was also found at a power station water outlet.) In the spring of 2016 the upper shore area at Archirondel was found to have dozens of *H. sanguineus* specimens on it. A public appeal and subsequent searching produced other hotspots from around the Jersey coast suggesting that the species is widespread, abundant and fully established. Limited searching in September 2016 suggests that *H. sanguineus* is still rare on Guernsey.

A student spent the summer of 2016 researching the population dynamics, ecology and behaviour of *H. sanguineus* on Jersey. Early results suggest that it is largely restricted to upper middle rocky shores where there is an abundance of loose stone and that, in this habitat, it is not competing directly with the native Green Shore Crab (*Carcinus maenas*). It was also concluded that the *H. sanguineus* population was still expanding

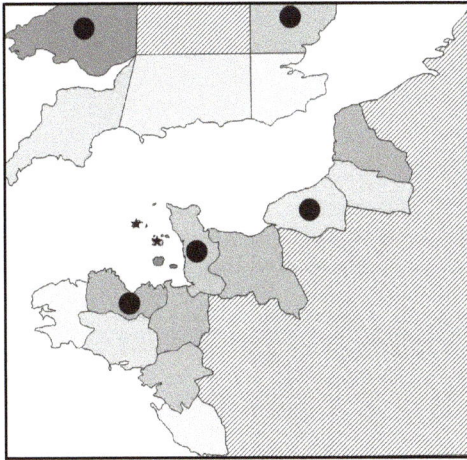

around the island's coastline. In confirmation of this, the crab was identified on Les Écréhous in July 2017 which represents its first report from one of the offshore reefs within the Channel Islands.

Its absence from other habitats and the lower shore will hopefully minimise the impact of *H. sanguineus* but the speed of its expansion mean the population will require close monitoring for several years.

Top left: *A specimen from Le Hurel (Jersey) showing the square carapace and patterning.* Top right: *A specimen at La Collette (Jersey).* Bottom left: *A specimen from Archirondel (Jersey) showing the claws.* Bottom right: *A crab from Archirondel (Jersey) in its natural habitat.*

# *Hemigrapsus takanoi*
## Crab

Threat score: 24

| ES | Hab | Tox | Econ |
|----|-----|-----|------|
| 3 | 4 | 1 | 2 |

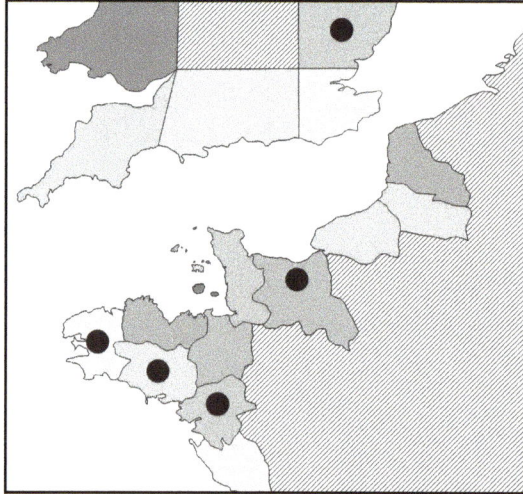

Native to the North-west Pacific, this crab was reported in La Rochelle in 1994 and has since spread north into Brittany and south into Biscay. However, it has also been transported to several other isolated locations including the eastern English Channel and Baltic Sea. The first British records were in 2014 when specimens were found on the Kent and Essex coasts. It is known from the north Brittany coast although it does not yet seem to have been found in Normandy or the Channel Islands.

This species has shown an ability to adapt to local conditions and to spread rapidly. Given that populations exist in the eastern English Channel and northern Brittany, its arrival in the Channel Islands in the near future seems probable. It is listed as being a high threat to the British Isles by Roy *et al.* (2014).

# *Rhithropanopeus harrisii*
## Harris Mud Crab

Threat score: 1

| ES | Hab | Tox | Econ |
|----|-----|-----|------|
| 1  | 1   | 1   | 1    |

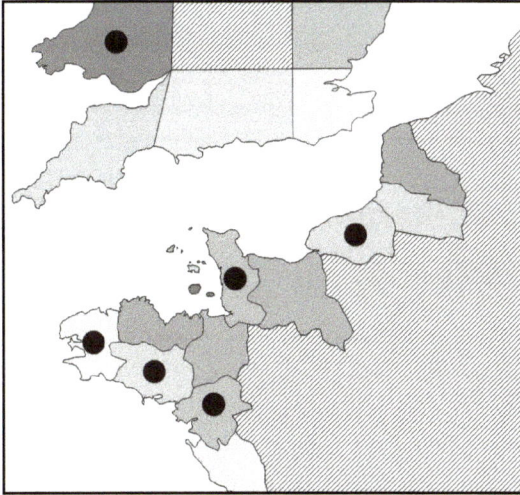

A North American small crab recorded from the north Normandy coast in the 1950s but which has since spread to many places along the Atlantic coast. There is only one record in the UK from Cardiff in 1996. Although it can tolerate fully marine conditions, this is a species that prefers sheltered lower salinity habitats such as estuaries and lagoons. It is unlikely to establish itself in the Channel Islands.

# *Pachygrapsus marmoratus*
## Marbled Shore Crab

| ES | Hab | Tox | Econ |
|----|-----|-----|------|
| 3 | 5 | 1 | 2 |

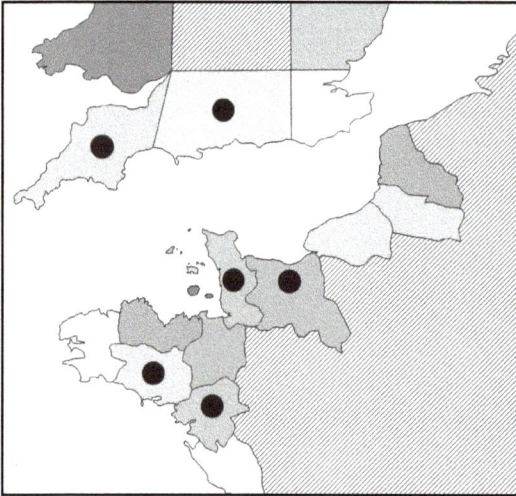

A small crab from the Mediterranean which has established populations in southern Brittany. Until recently it was assumed that the northern breeding limit of *P. marmoratus* was the Mobihan region of Brittany. However, in 1996 specimens were found in Southampton, in 2006 in Germany and in 2008 on the western coast of Normandy. This movement is sometimes explained as being due to rising regional sea temperatures.

Minchin *et al.* (2013) stated that a specimen had been found in Jersey in 2009 but no evidence of this can be found and it is possible that this is a reference to the discovery of *Hemigrapsus sanguineus* that year. *P. marmoratus* shares much the same middle to upper shore habitat of *H. sanguineus* and intensive field studies of the latter species in 2016 produced no records for *P. marmoratus* suggesting that it is not yet established on Jersey.

*P. marmoratus* is known to breed prolifically and the principal threat is probably to other crab species sharing the same ecological niche. At present this would primarily mean the native Green Shore Crab (*Carcinus maenus*) and the recently established Asian shore crab (*H. sanguineus*). Recent regional records suggest that is possible that *P. marmoratus* could establish itself within the Channel Islands in the near future.

# *Gibbula albida*
## White Topshell

Threat score: 6

| ES | Hab | Tox | Econ |
|----|-----|-----|------|
| 2 | 3 | 1 | 1 |

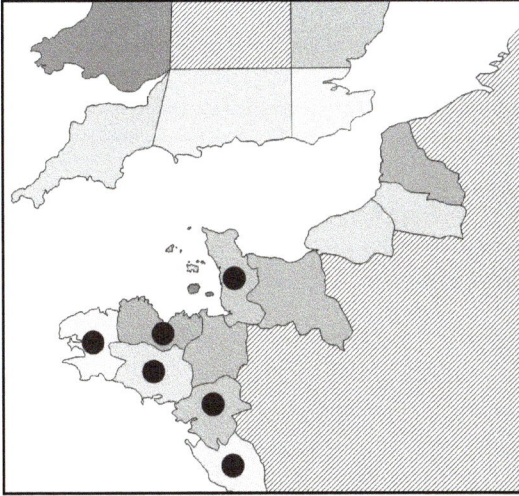

A Mediterranean species that was accidentally imported into the southern Brittany oyster region in the 1980s but which has since been reported in north Brittany and Normandy. Its spread may be linked to the movement of aquaculture stock and its occurrence on adjacent coasts suggests that it may establish itself in the Channel Islands although targetted searches on Jersey have not yet yielded any specimens. It can compete with a native species, the Turban Topshell (*Gibbula magus*), but the overall threat is considered to be low.

# Crepidula fornicata
## American Slipper Limpet

Threat score: 125

| ES | Hab | Tox | Econ |
|----|-----|-----|------|
| 5 | 5 | 1 | 5 |

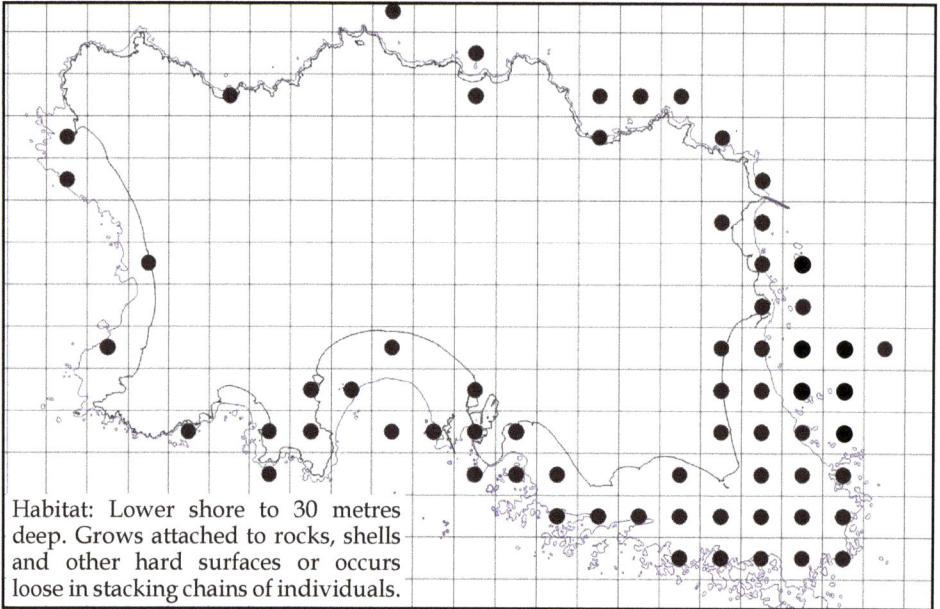

Habitat: Lower shore to 30 metres deep. Grows attached to rocks, shells and other hard surfaces or occurs loose in stacking chains of individuals.

The Slipper Limpet presents probably the greatest threat to the marine environment within the Normano-Breton Gulf. It has been introduced into Europe from the USA on at least two occasions; the first into the UK in the nineteenth century and the second into France in the 1970s. On both occasions the importation of oyster seed was the transport vector. Once introduced, the Slipper Limpet spread rapidly to many other areas via the movement of aquaculture stock, ships' hulls, etc.

The first record of Slipper Limpets in the Normano-Breton Gulf come from the Brest area of Brittany in 1949 but it is not until the mid-1970s that it started to occur in the Channel Islands area probably as a consequence of the movement of aquaculture stock. The first Channel Islands records were in 1975 from the area of seabed just north of Les Écréhous (Rètiere, 1979) and by the 1980s it was common on the seashore. A seabed survey in 1996 found large numbers of Slipper Limpets in the Bay of Granville region including off the east and south coasts of Jersey. The biomass of Slipper Limpets in 1996 was estimated at 107 tonnes with seabed coverage reaching 100% in some areas. This raised serious concerns about the effect that Slipper Limpets were having on the local environment. The seabed area was resurveyed in 2004 revealing the density and coverage had increased three-fold and the biomass had risen to 149 tonnes (Blanchard, 2009).

By this time the Slipper Limpet had begun to impact severely on the aquaculture industry within the Bay of Mont St Michel and there were

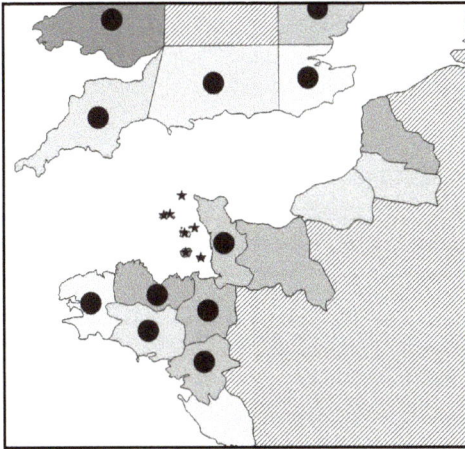

suggestions that this was affecting offshore scallop dredging. Areas of seabed with a density of Slipper Limpets >50% were found to be functionally barren. Within the Bay of Mont St Michel around 25 km² was in this state, affecting aquaculture and all fishing methodologies.

Recent surveys using video cameras have revealed large areas of seabed off Jersey's east and south coasts where Slipper Limpet coverage is >50%. There is currently serious concern about the rapid geographic spread and increasing density of Slipper Limpets in Jersey's shallow marine zone and the potential severe environmental and economic threats.

The Slipper Limpet situation requires urgent assessment and study to quantify the problem, predict its ultimate and decide what actions can be taken to mitigate the environmental and social impacts.

TL: *A stacking of chain of Slipper Limpets at La Collette, Jersey.* TR: *Slipper Limpets attached to a rock in Grouville, Jersey.* BL: *Seabed off the east coast of Jersey with 100% coverage of Slipper Limpets.* BR: *Slipper Limpets attached to Great Scallop and Dog Cockle shells dredged off Jersey's east coast.*

# *Potamopyrgus antipodarum*
## Jenkin's Spire Shell

Threat score: 4

| ES | Hab | Tox | Econ |
|----|-----|-----|------|
| 2 | 2 | 1 | 1 |

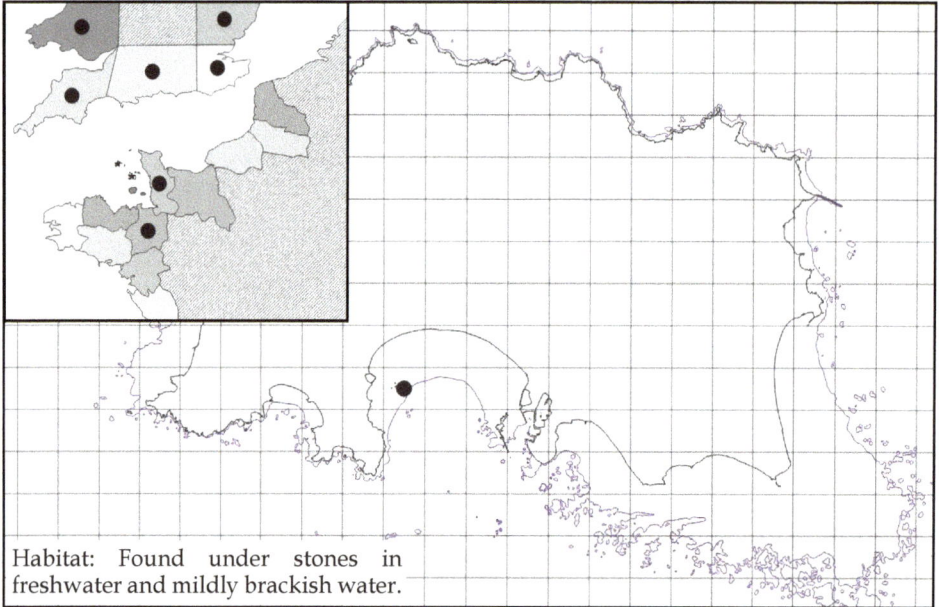

Habitat: Found under stones in freshwater and mildly brackish water.

First recorded in the UK in 1859 having probably arrived inside barrels of drinking water carried from Australia. Although primarily a freshwater species, Jenkin's Spire Shell can survive in brackish water and empty shells have been discovered inside St Aubin's Harbour although whether this is indicative of actual seashore inhabitation is questionable. A lack of true brackish water habitats makes it unlikely that this species will colonise Jersey's marine environment.

# *Fusinus rostratus*
## Gastropod

Threat score: 9

| ES | Hab | Tox | Econ |
|----|-----|-----|------|
| 3 | 3 | 1 | 1 |

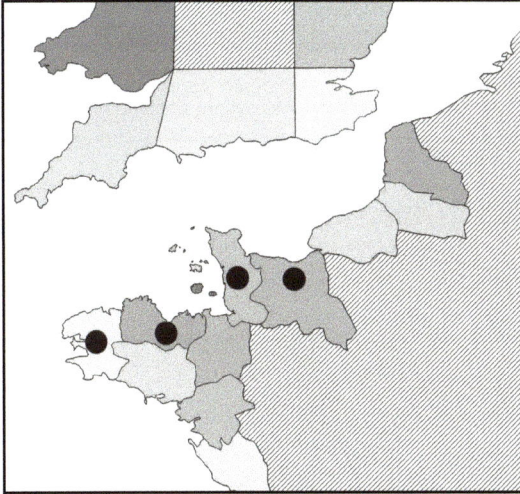

Native to the Mediterranean but accidentally introduced into southern Brittany via aquaculture. By 2007 specimens were being reported from shallow marine areas off the north Brittany coast and Normandy. *F. rostratus* seems to be associated with areas of maerl and it seems probable that specimens will eventually be found in Channel Island waters. The threat this species presents is as yet undefined; it is a large animal that feeds primarily on polychaetes but its association with fragile habitats such as maerl is of concern.

# Tritia neritea
## Gastropod

| ES | Hab | Tox | Econ |
|----|-----|-----|------|
| 1 | 1 | 1 | 1 |

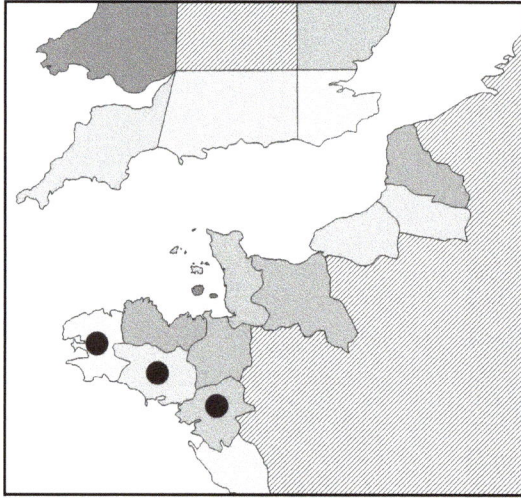

This gastropod mollusc is native to the Mediterranean and southern Iberian peninsula but in 1984 it became established in southern Brittany, probably through aquaculture. It remains common in southern Brittany but has not spread further north than the Loire-Atlantic coast. It is not considered to be a threat to the Normano-Breton Gulf and its preference for sheltered muddy situations and temperate water preferences may limit its ability to establish itself in the Channel Islands.

# Ocenebra inornata
## Japanese Sting Winkle

Threat score: 24

| ES | Hab | Tox | Econ |
|---|---|---|---|
| 2 | 3 | 1 | 4 |

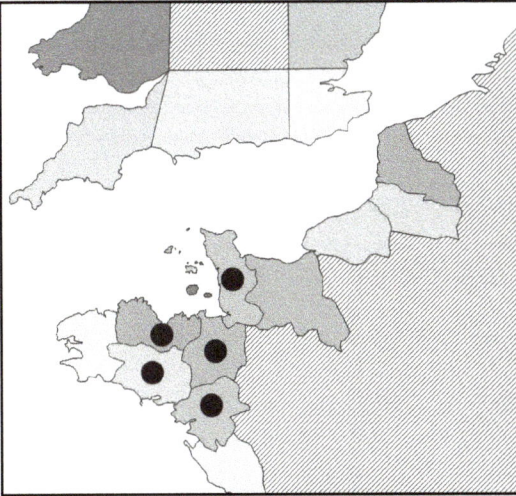

Native to the North-west Pacific, the first European discovery of *O. inornata* was in the Charente-Maritime region of France in 1995 from where it spread quickly north into Brittany. In 2003 it was reported from the Bay of St Malo and then in Denmark shortly afterwards.

As a serious pest to oyster beds, the spread of *O. inornata* is of concern with the species being listed as a high threat to the British Isles by Roy *et al.* (2014). It is particularly associated with *Crassostrea gigas* and the movement of aquaculture stock (on which eggs are often laid) may be a vector between distant locations. There is a strong similarity between *O. inornata* and the native Oyster Drill (*Ocenebra erinacea*) with many European specimens being initially misidentified. Given its presence on the adjacent Normandy coast, there is a possibility that *O. inornata* will become established in the Channel Islands either through natural spread or perhaps via aquaculture. It is suspected to be a competitor to some native species and presents an economic threat to the aquaculture industry.

# *Rapana venosa*
## Veined Rapa Whelk

Threat score: 80

| ES | Hab | Tox | Econ |
|----|-----|-----|------|
| 4 | 4 | 1 | 5 |

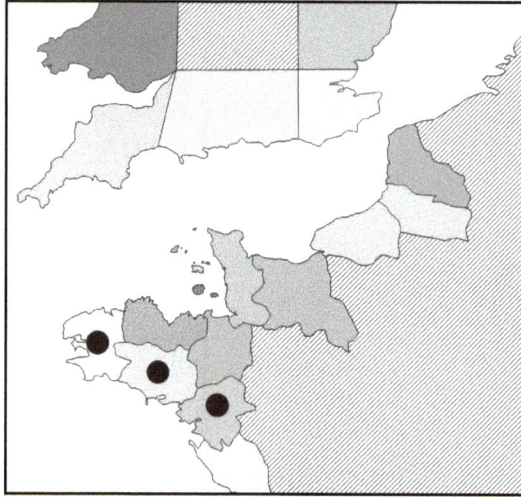

Native to the North-west Pacific, *R. venosa* is a large (18 cm) predatory gastropod which entered Europe in the 1950s via the Black Sea. In 1998 specimens were found in southern Brittany where monitoring suggested that the population was small, stable and not spreading rapidly. In 2005 specimens were dredged from two locations in the North Sea suggesting that it may be present offshore elsewhere in Europe, including the English Channel.

As a large predatory gastropod, there are concerns that *R. venosa* may consume ecologically and commercially important species, such as oysters. The species is considered to be a potential threat to the English Channel with a prediction that the 'impacts could be enormous'. It is regarded as a high threat to the British Isles by Roy *et al.* (2014) and with apparent population centres in the North Sea and southern Brittany, *R. venosa* has the potential to reach the Channel Islands although the timescale for this is hard to predict. It should be regarded as a potential threat locally and it could perhaps be part of an awareness campaign amongst fishermen, naturalists and divers.

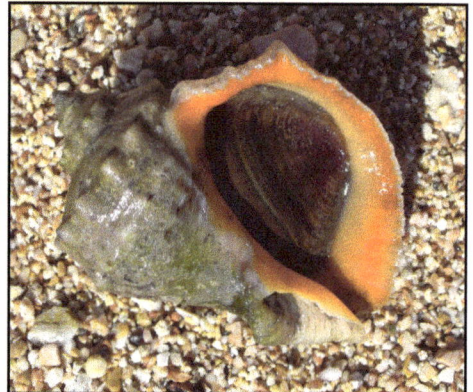

Above and right: *Examples of* Rapana venosa *from the Black Sea region.*

# *Urosalpinx cinerea*
## American Sting Winkle

Threat score: 8

| ES | Hab | Tox | Econ |
|----|-----|-----|------|
| 1 | 1 | 2 | 4 |

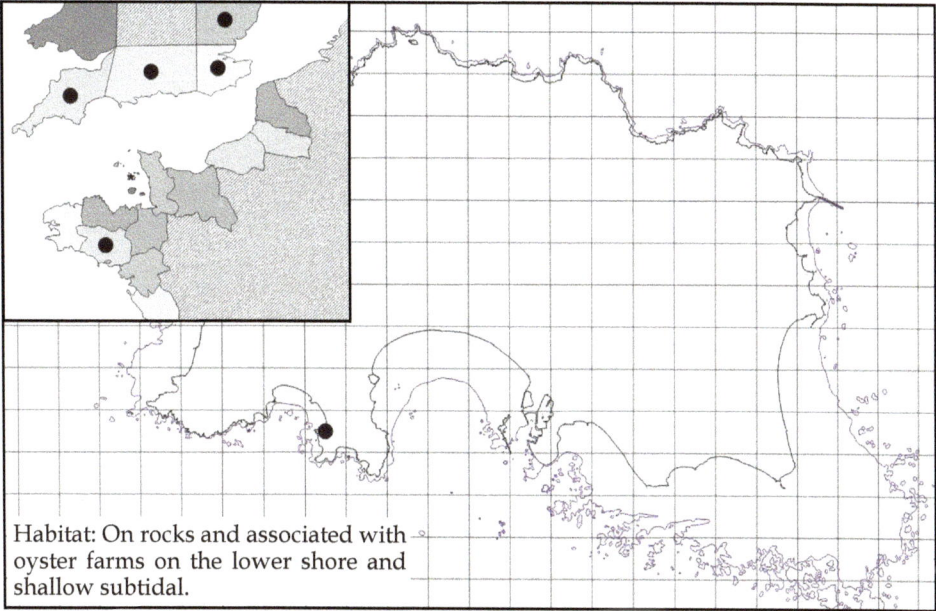

Habitat: On rocks and associated with oyster farms on the lower shore and shallow subtidal.

First recorded in Essex in 1927 after having been introduced from the USA with oyster stock. It does not spread rapidly and is usually closely associated with oyster aquaculture sites where it can cause serious damage. Beyond a lone record from Portelet Bay in 1983, the American Sting Winkle is not known from the southern English Channel. Based on this it is probable that the Jersey record is a misidentification, probably of an abnormal specimen of a Dog Whelk (*Nucella lapillus*), other examples of which have been reported to the Société Jersiaise. If *U. cinerea* ever was resident in Jersey then it has since become extinct.

# *Haminoea japonica*
## Japanese Bubble Snail

Threat score: 24

| ES | Hab | Tox | Econ |
|----|-----|-----|------|
| 2  | 2   | 3   | 2    |

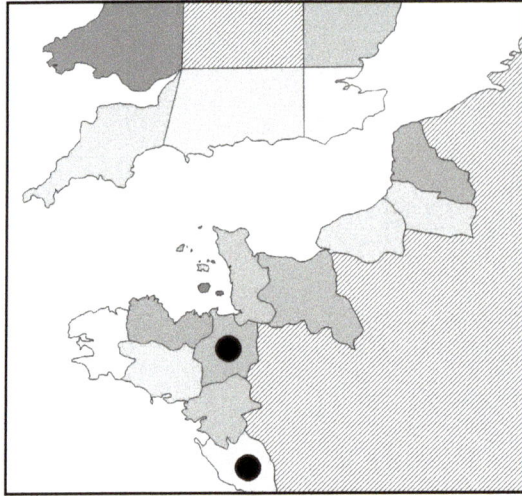

Native to the North-west Pacific *H. japonica* has been reported from North America and Europe. Initially reported from Venice in 1992 and then from other locations in the Mediterranean and Bay of Biscay, *H. japonica* is believed to have been transported with aquaculture stock to the Bay of St Malo in 2003.

The animal has a distinctive internal shell (arrowed below) and prefers sheltered muddy habitats, which tend to be uncommon in the Channel Islands. It seems to be transported mainly with aquaculture stock which, with existing biosecurity measures, should lessen its chances of reaching the Channel Islands. It has been associated with human health problems (specifically the condition 'Swimmers' Itch') in Calfornia due to parasitic trematodes to which it can act as a host.

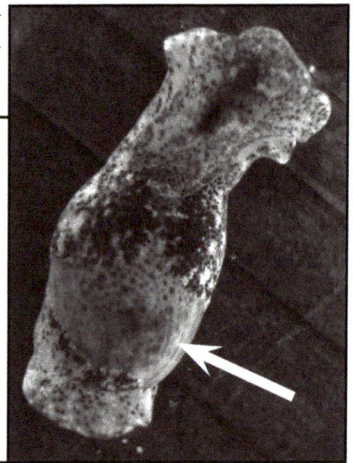

*Above: A specimen from California.*
*Right: The glassy internal shell is arrowed.*

# *Mytilopsis leucophaeata*
## Conrad's False Mussel

Threat score: 1

| ES | Hab | Tox | Econ |
|----|-----|-----|------|
| 1 | 1 | 1 | 1 |

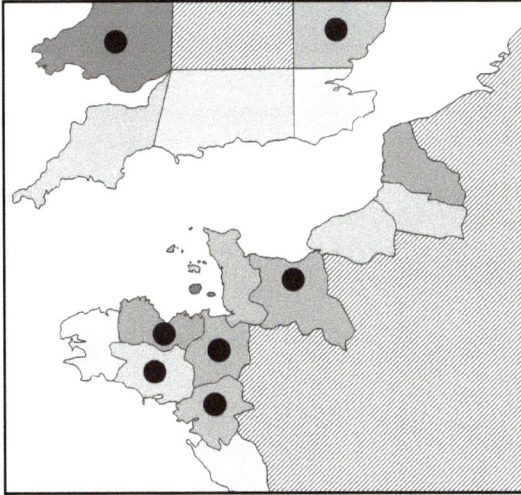

A small species of mussel from the Gulf of Mexico that was first reported in Brittany in 1970. It is noted as a fouling organism and as a species that can outcompete native mussels. Although reported from several regional locations, *M. lecuophaeata* does not grow in fully marine conditions and is unlikely to be found in the Channel Islands.

2mm

# Crassostrea gigas
## Pacific Oyster

Threat score: 72

| ES | Hab | Tox | Econ |
|----|-----|-----|------|
| 3  | 4   | 3   | 2    |

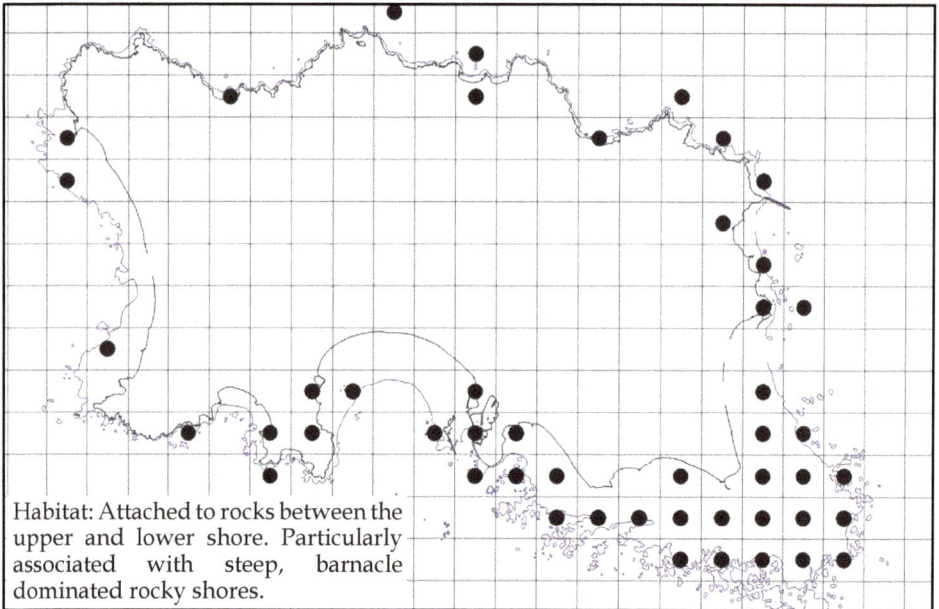

Habitat: Attached to rocks between the upper and lower shore. Particularly associated with steep, barnacle dominated rocky shores.

Introduced into Essex in 1927 as an aquaculture species, the Pacific Oyster has adapted well to local seas with wild breeding stocks being reported from 1970 onwards. Spread has been rapid and this species occurs across northern Europe.

The first Channel Islands report of wild specimens was from Jersey in 1982 but by the 1990s they had become common on rocky shores across the islands. It is probable this wild stock originated either from aquaculture concessions on Jersey or from similar ventures on the adjacent French coast. Currently *C. gigas* may be found on rocky shores across the islands where it may be one of the commonest large molluscs. It is also common on offshore reefs such as Les Écréhous which suggests that larvae are able to move around the local area with relative ease.

*C. gigas* is common within the Channel Islands but has not reached the densities seen in some parts of Brittany where the species has started to form artifical reefs on some shores, smothering other marine life. Also, in some rocky areas the aggregation of cemented lower valves on intertidal rocks has produced a distinctive white band. The ability of the oysters to colonise artificial structures, such as pilings, piers and aquaculture equipment, has led to them being declared a menace in some parts of Brittany. It is, however, recognised that there is little that can be done to stop the oysters from settling and growing

Only in a few places within the Channel Islands (such as behind St Aubin's Fort) has the colonisation of oyster shells become noticeably dense

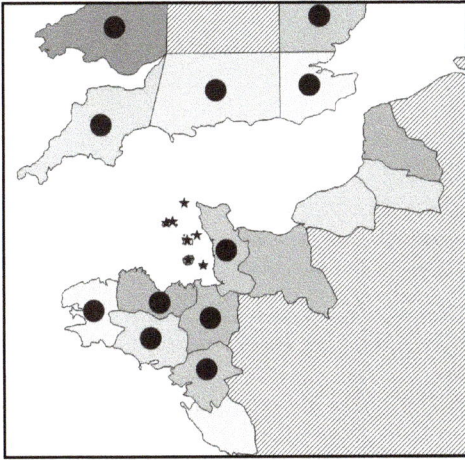

but this is nowhere near the levels seen in southern Brittany where entire rock surfaces are covered by thousands of individuals.

The Pacific Oyster is at present considered to be a medium to high level threat to the environment. This is an obvious seashore species whose density could be monitored using photography or simple survey techniques. The species is sometimes targeted by low water fishermen.

*Top left: and right: Naturalised* **Crassostrea** *specimens growing on rocks on Jersey's south coast. Bottom left: Flat and Pacific oysters on sale in Brittany. Bottom right: A bank of discarded oyster shells near to an aquaculture area in the Bay of Mont St Michel.*

# *Choromytilus chorus*
## Chorus Mussel

Threat score: 1

| ES | Hab | Tox | Econ |
|----|-----|-----|------|
| 1 | 1 | 1 | 1 |

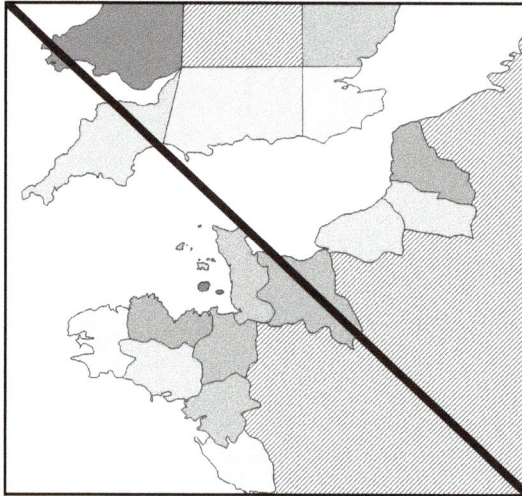

A very large species of mussel from the SW Atlantic that has been cultivated commercially in some regions. Listed as being a British species by Minchin *et al.* (2014), additional details have been difficult to find. As *C. chorus* is described as being not established in British waters and does not appear on standard non-native European lists, it is probably not spreading rapidly and does not present a serious threat. Probably not a species that will be present in Channel Island waters in the foreseeable future.

# *Mizuhopecten yessoensis*
## Scallop

Threat score: 1

| ES | Hab | Tox | Econ |
|----|-----|-----|------|
| 1 | 1 | 1 | 1 |

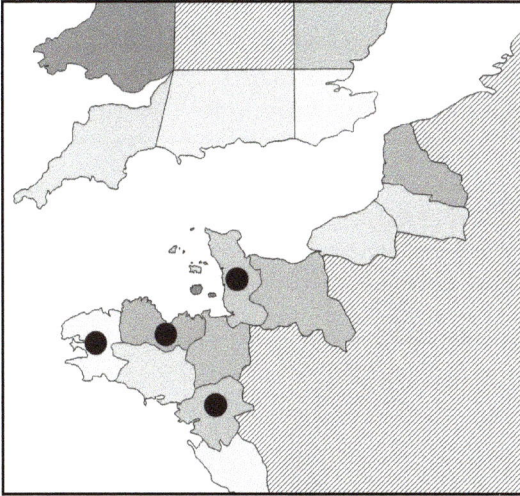

This species of scallop was deliberately introduced into northern Brittany in the 1970s as a farmed species. It does not appear to have reproduced or spread and the prospect of it being established the Channel Islands is at present remote.

# *Ensis directus*
## Jack Knife Clam

Threat score: 12

| ES | Hab | Tox | Econ |
|----|-----|-----|------|
| 3 | 4 | 1 | 1 |

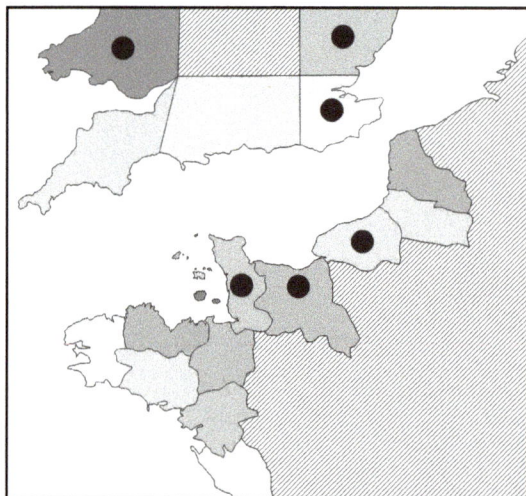

Introduced from North America in the 1980s, this species is common in southern England and the north Normandy coast. It has not, however, been reported from the Manche coast or Brittany and targeted searches on Jersey have not yielded specimens. *E. directus* is very similar to native razor clam species such as *E. arcuatus* and *E. ensis* both of which it can out-compete. Identification of *E. directus* is difficult because of its similarity to *E. arcuatus*. The main impact seems to be on native species and the spread of *E. directus* along the Normandy coast should be monitored and periodic checks made on jersey as this species has the potential to spread into Channel Island waters.

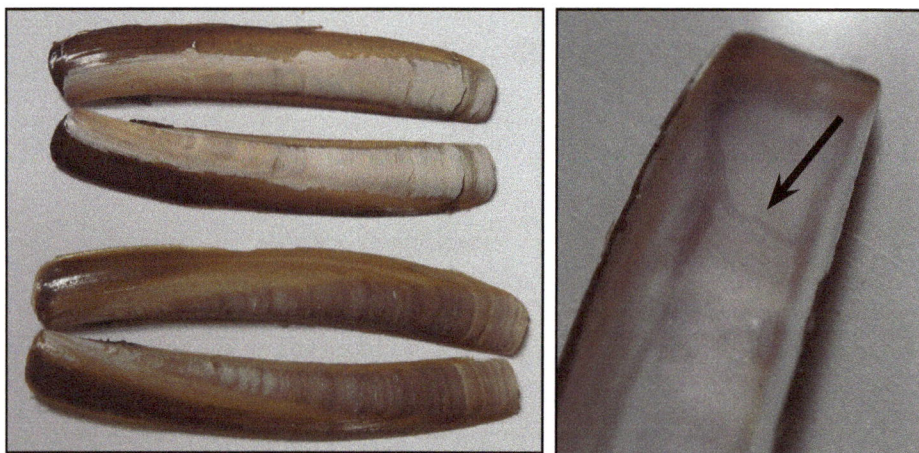

Left: *A specimen of* Ensis directus *(top shell) next to* Ensis arcuatus. Right: *The pallial sinus in* E. directus *has a bend or kink in it (arrowed) that* E. arcuatus *lacks.*

# *Mercenaria mercenaria*
## Hard Clam

Threat score: 4

| ES | Hab | Tox | Econ |
|----|-----|-----|------|
| 2 | 2 | 1 | 1 |

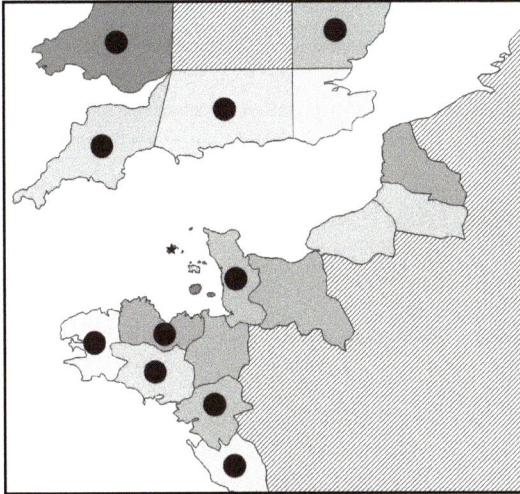

A commercial species from North America whose only native Channel Island record is a single live specimen from Guernsey but this has been questioned. A worn single valve was recovered near to Green Island, Jersey, in February 2017. No licences have ever been issued for the importation of *M. mercenaria* into Jersey and so the origin of this specimen is unresolved.

# *Ruditapes philippinarum*
## Manilla Clam

Threat score: 32

| ES | Hab | Tox | Econ |
|----|-----|-----|------|
| 2 | 4 | 2 | 2 |

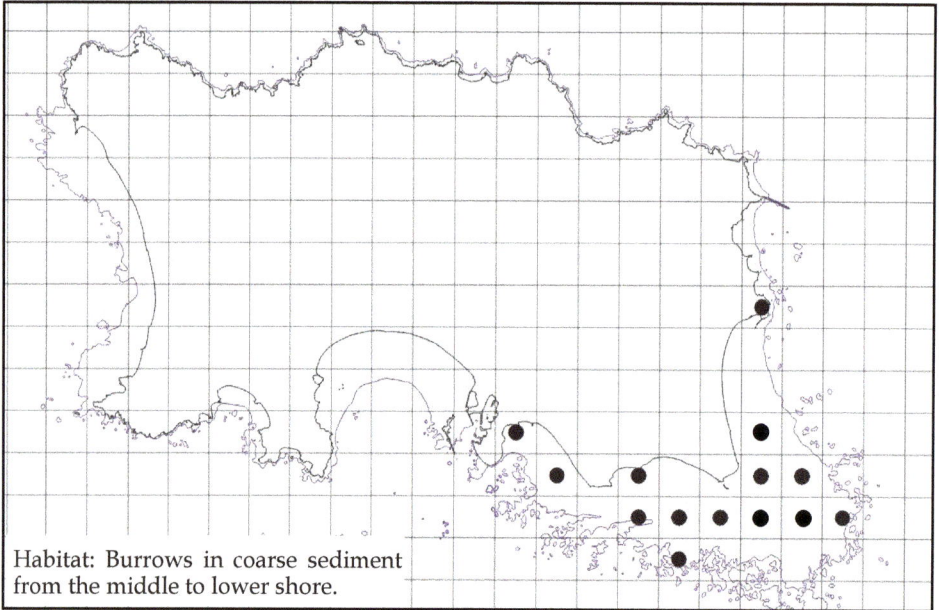

Habitat: Burrows in coarse sediment from the middle to lower shore.

The Manilla Clam was introduced into Europe in the 1970s from the North-west Pacific Ocean as a new species for the aquaculture industry. It was initially cultivated in Normandy and Brittany from where seed stock was exported to many other places in Europe, including the Channel Islands. The Manila Clam is suited to European seas and is able to breed freely, leading to the establishment of localised wild populations in those areas where it has been cultivated. The movement of stock is also believed to have been responsible for the dispersal of several other non-native species such as the Japanese Bubble Snail (*Haminoea japonica*).

The species was introduced into Jersey in 1986 for aquaculture purposes when an experimental concession was seeded in St Catherine's Bay (Fig. 1.2). The following year plans were made to sow up to eight million *R. philippinarum* seed but how much of this was actually achieved is unknown. However, it would appear that the attempt at farming this species was relatively short-lived and had certainly ceased by 2000. It is thought that some Manila Clam beds were temporarily in place near to La Rocque and Le Hocq but that a high mortality rate due to crab predation made the venture unviable.

Monitoring since 2009 has found *R. philippinarum* to be common along on Jersey's east and south coasts and in some areas the empty shells may be abundant. The pattern of distribution (which is restricted to the south and east coasts) suggests that the island's wild Manilla Clam population originates from the attempts at farming this species during the late 1980s

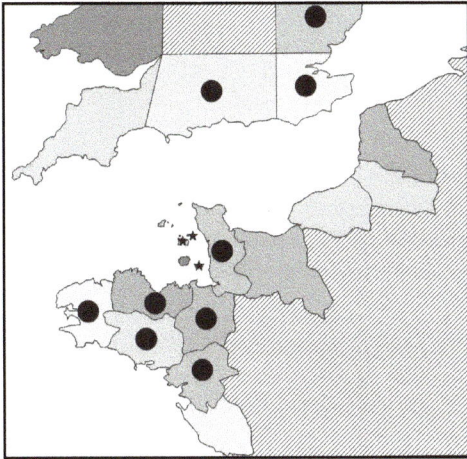

and 90s. However, the species has been also been widely cultivated along the Normandy and Brittany coasts and this cannot be dismissed as the source of Jersey's Manilla Clam population.

The Manilla Clam generally restricts itself to middle and upper lower shore coarse sediment environments and as such does not appear to compete directly with native *Tapes/Venerupis* species which are generally found lower down on the shore. Its spread is liable to continue but based on present knowledge it presents a low to medium level threat to the local environment. Given that it can breed in local waters, it is recommended that no further attempts should be made to farm this species in order to prevent further risk to local habitats and species.

*Specimens of* R. philippinarum *from Jersey's south-east coast. The distinctive interior colouration is arrowed.*

# *Mya arenaria*
## Sand Gaper

Threat score: 2

| ES | Hab | Tox | Econ |
|----|-----|-----|------|
| 2 | 1 | 1 | 1 |

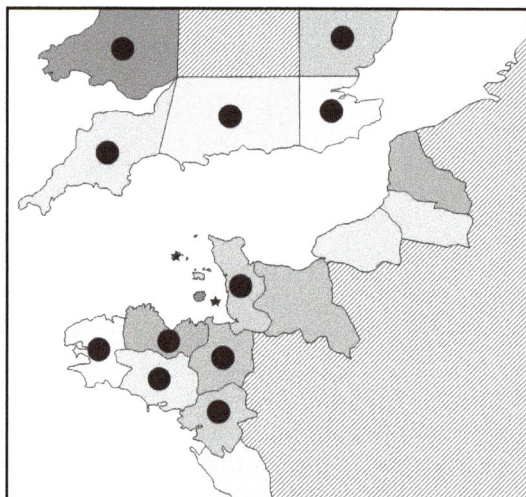

A cryptogenic species that may have been imported into this part of Europe during historical times. The only Channel Island record is a single dead valve from Guernsey. *M. arenaria* is unlikely to have been established here and in many parts of northern Europe is considered to be a native species.

# *Lyrodus pedicellatus*
## Blacktip Shipworm

Threat score: 3

| ES | Hab | Tox | Econ |
|----|-----|-----|------|
| 1  | 1   | 1   | 3    |

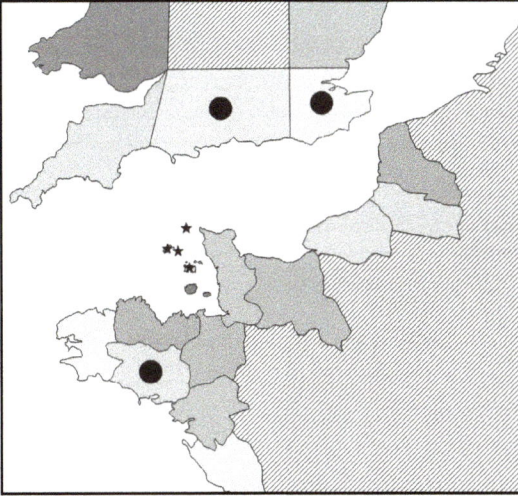

A cryptogenic species with a wide global distribution. It has a number of records from the Channel Islands, most of them historical, but an absence of wooden piers has probably rendered this a very rare species. The Normano-Breton Gulf is generally considered to be at the northern edge of its range although a handful of records are known from southern England. Although its burrowing can weaken and undermine wooden structures it is not common enough to present a threat in the islands.

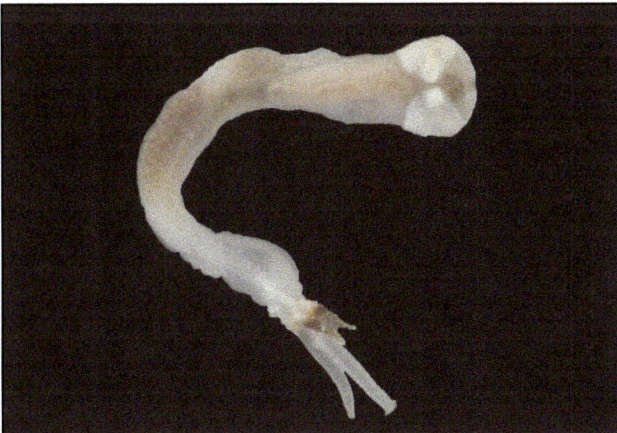

# *Teredo navalis*
## Shipworm

Threat score: 3

| ES | Hab | Tox | Econ |
|----|-----|-----|------|
| 1  | 1   | 1   | 3    |

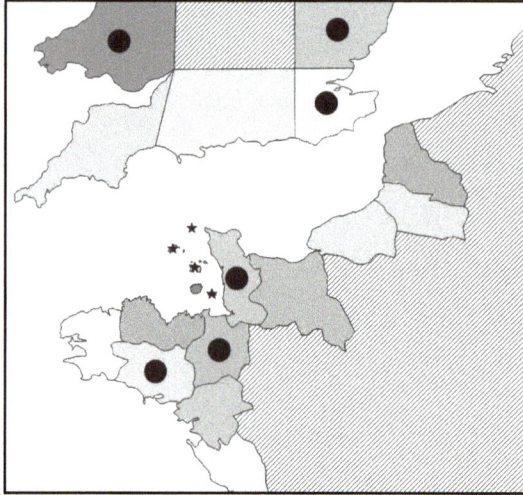

A cryptogenic species with an extensive history of recording in northern Europe. For centuries the Shipworm was a menace to the hulls of ocean going vessels but anti-fouling and rigid hulls has lessened this problem. Specimens still arrive in driftwood but the species has been in the region for such a long time that it is often treated as being native.

*Left: The characteristic tube of a shipworn in driftwood foudn on Jersey in 2009.*
*Right: The a shipworm valve recovered from the same piece of driftwood.*

# *Bugula neritina*
## Bryozoan

Threat score: 27

| ES | Hab | Tox | Econ |
|----|-----|-----|------|
| 3 | 3 | 1 | 3 |

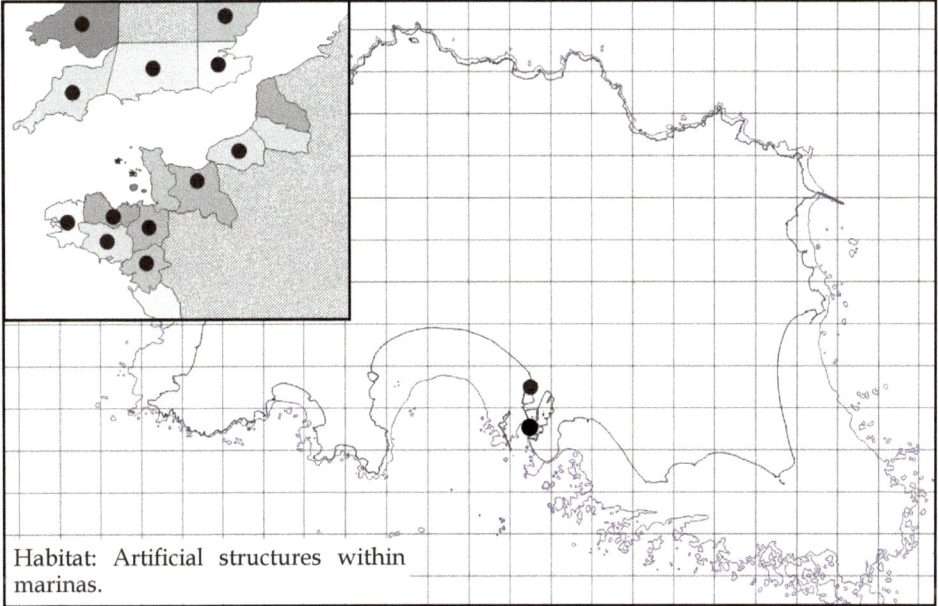

Habitat: Artificial structures within marinas.

The history of this species in Europe is complex. First identified in the 1911 in Plymouth, *Bugula neritina* was afterwards reported from southern England and France but was for a short while in the 1990s declared extinct in the UK. The UK presence was re-established in 2004 and the first Channel Island report came from Guernsey marina in 2007. It was probably also established in Jersey's marinas at this time although this was not confirmed until 2014. Although *B. neritina* continues to spread in Europe, it primarily remains a species that is associated with ports and marinas where it has sometimes been reported as a fouling organism. This has not been observed in Jersey where and it is considered to be a low level threat to the island's wider marine environment.

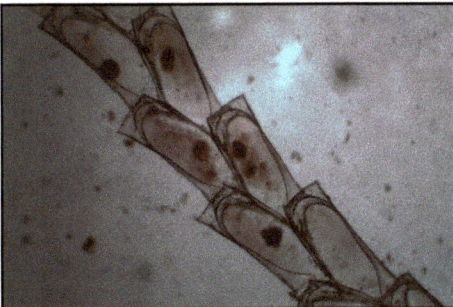

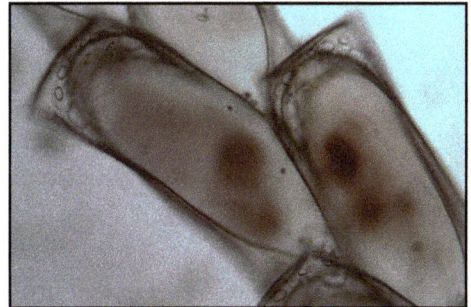

Left: *A specimen of* B. neritina *taken from St Helier Marina in 2016 magnified x20.* Right: *Details of the zooids on the same specimen.*

# Bugula stolonifera
## Bryozoan

Threat score: 27

| ES | Hab | Tox | Econ |
|----|-----|-----|------|
| 3  | 3   | 1   | 3    |

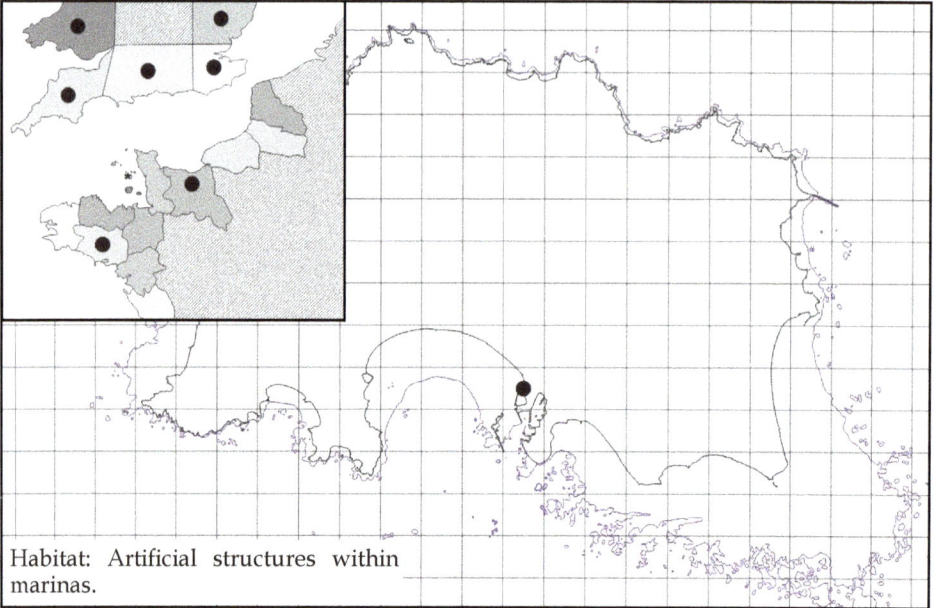

Habitat: Artificial structures within marinas.

Possibly recorded from the nineteenth and early twentieth centuries in the UK, the first modern records of *B. stolonifera* were not until the 1970s from eastern English Channel ports. It has not spread rapidly in northern Europe and may be restricted to sheltered marinas and harbours. Specimens were first found in St Helier's Elizabeth marina in 2009 but it has not so far been identified on the wider seashore. It is not common and presents a low level threat to the local environment. In 2016 a deliberate search of St Helier's marinas did not reveal any specimens.

*Left and right: Specimens of* **B. stolonifera** *found at Elizabeth Marina, Jersey in 2009.*

# *Caulibugula zanzibarensis*
## Bryozoan

Threat score: 4

| ES | Hab | Tox | Econ |
|----|-----|-----|------|
| 2  | 2   | 1   | 1    |

Arcachon

Discovered living offshore from Le Bassin d'Arcachon (Biscay) in 2003, this cheilostome bryozoan has the potential to spread northwards into Brittany. Although noted as a potential fouling species, it is not regarded as a serious threat and is unlikely to be found in the Channel Islands in the short or medium term.

# *Tricellaria inopinata*
## Bryozoan

Threat score: 18

| ES | Hab | Tox | Econ |
|----|-----|-----|------|
| 2 | 2 | 1 | 3 |

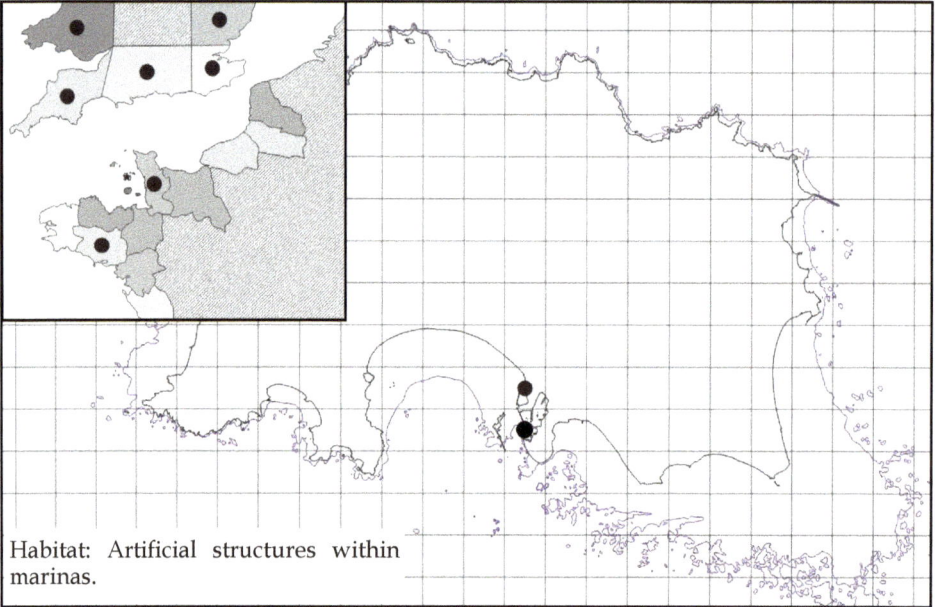

Habitat: Artificial structures within marinas.

A native of the North-east Pacific, *T. inopinata* was first reported in Europe in 1982 from Venice but by the 1990s it had reached the English Channel and was reported from Brittany and Normandy in 2001 and 2003. Noted as a competitor with native species, *T. inopinata* is almost exclusively associated with marinas and deliberate searches in the UK have resulted in many new sites being discovered. A search of St Helier Harbour in 2016 revealed *T. inopinata* to be abundant on pontoons at La Collette and St Helier Marinas.

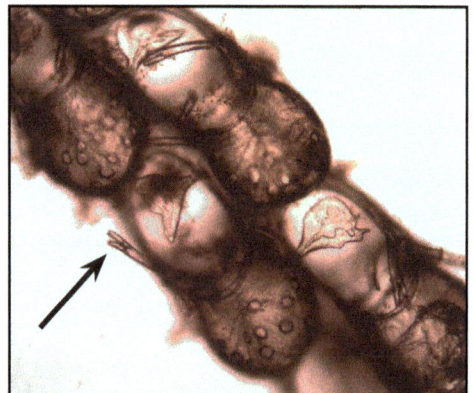

*Microscopic detail of a colonies from St Helier Marina, Jersey. A distinctive bifid spine is arrowed.*

# *Schizoporella errata*
## Bryozoan

Threat score: 27

| ES | Hab | Tox | Econ |
|----|-----|-----|------|
| 3 | 3 | 1 | 3 |

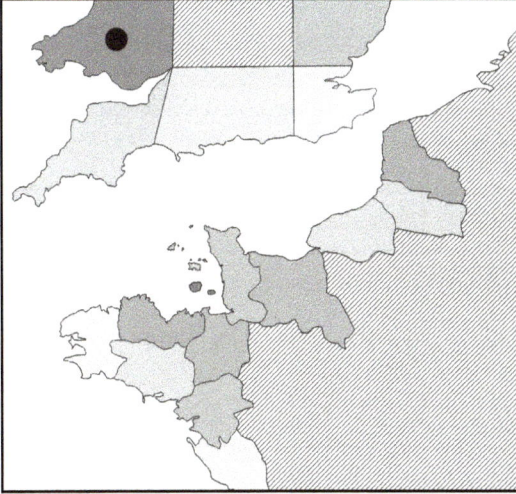

A colonial bryozoan that is native to the Mediterranean but which has spread to many other parts of the globe including the UK where specimens were reported in the 1970s. The principal issues with *S. errata* are fouling and competition with native species. Individual colonies can be large and form encrustations up to 25 cm in height on buoys, ships and piers. This, and its ability to compete for space, has led to it being listed as a medium threat to the British Isles by Roy et al. (2014). *S. errata* generally prefers shallow marine (0 to 10m) habitats and is particularly prevalent in harbours. It has spread to many temperate seas around the world and, while not currently in the English Channel, it is thought that the species has the potential to spread there and become established in the wider marine environment.

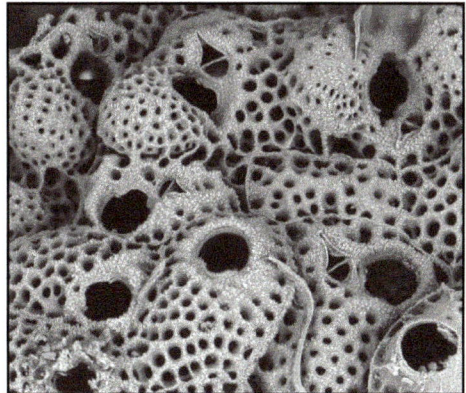

# *Watersipora subatra*
## Asian Bryozoan (=*W. subtorquata*)

Threat score: 24

| ES | Hab | Tox | Econ |
|----|-----|-----|------|
| 3 | 4 | 1 | 2 |

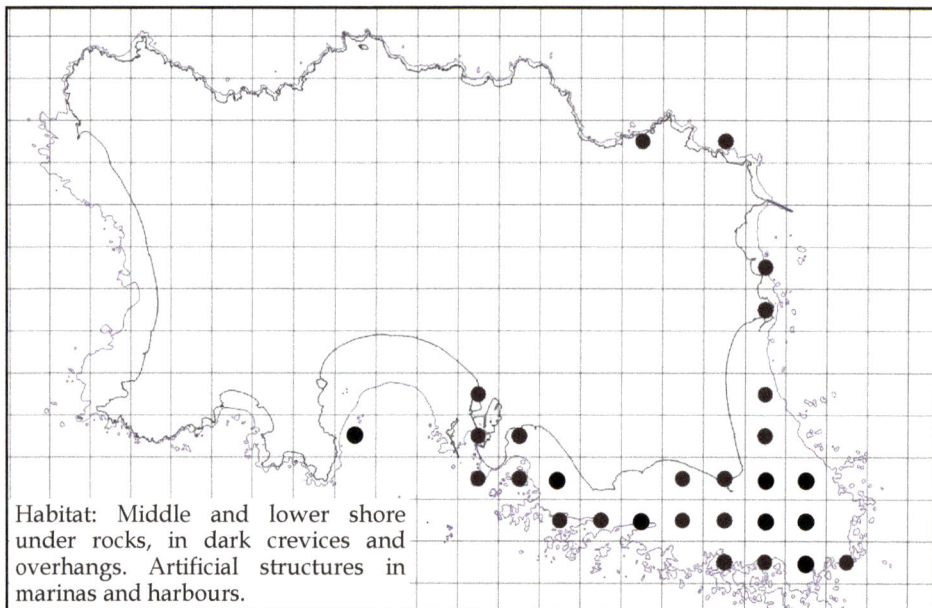

Habitat: Middle and lower shore under rocks, in dark crevices and overhangs. Artificial structures in marinas and harbours.

There remains some taxonomic confusion around the species name for this animal and, at the time of writing, it is still commonly referred to as *Watersipora subtorquata* (but see page opposite). First recorded on the Atlantic coast of France in 1983, it is possible that *W. subatra* arrived up to a decade earlier with oyster stock from Japan. It has spread slowly but consistently and was first identified in Guernsey in 2007, then Plymouth (UK) in 2008 and Jersey in 2009. In this respect it is one of those northerly-spreading species whose first British record is in the Channel Islands followed swiftly by the first UK specimens.

For several years *W. subatra* was restricted to the marinas in Jersey and Guernsey and it was not until February 2011 that the first specimen was discovered at La Collette, Jersey. Thereafter it started to spread although a majority of recorded specimens are on the south and east coasts and it is rare on the west coast. It is also common Guernsey's seashore.

As well as being geographically widespread, the abundance of *W. subatra* has increased significantly in recent years making it a common find under stones. In some situations, such as in deep crevices, in lower shore caves and overhangs, *W. subatra* can form 100% surface cover, sometimes over several square metres. This has also been observed underneath seasonal pontoons used in outlying harbours. Aside from demonstrating an aversion to bright situations, this suggests that *W. subatra* may be a pioneering species that can colonise vacant surfaces quickly.

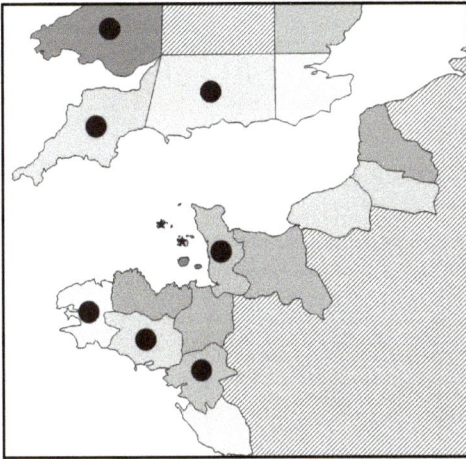

Although noted as a fouling organism, the main threat posed by *W. subatra* is its ability to outcompete native bryozoan species (principally *Schizoporella unicornis* and *Escharoides coccinea*) and to occupy surfaces to the detriment of other encrusting plants and animals. *W. subatra* is a threat to local biodiversity and it needs to be monitored closely. It appears to be intertidal in its range, and is absent or very rare from local kelp forests and some of the more delicate subtidal rocky habitats.

## Taxonomic Confusion

The genus *Watersipora* has suffered from much taxonomic confusion. Until recently *W. subatra* specimens found in the Normano-Breton Gulf were referred to as *W. subtorquata*. A recent review suggests that *W. subtorquata* specimens from the English Channel are actually the Japanese species *W. subatra*. This is also the case for bryozoan specimens identified in Normandy (Granville) as *W. subovoidea* in 2002. See Vieira *et al.* (2014) for more details.

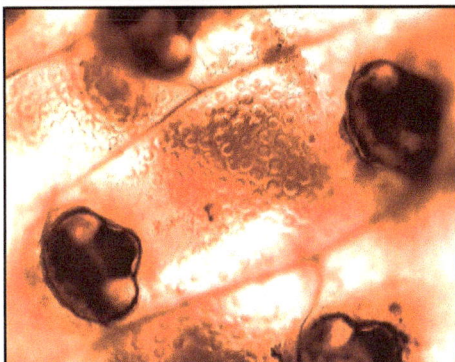

*TL:* **W. subatra** *under a rock. TR: Detail of the same colony. BL: Microscopic detail. BR: A pontoon at Gorey, Jersey, overgrown with* **W. subatra***.*

# Schizoporella japonica
## Orange Ripple Bryozoan

Threat score: 36

| ES | Hab | Tox | Econ |
|----|-----|-----|------|
| 3 | 4 | 1 | 3 |

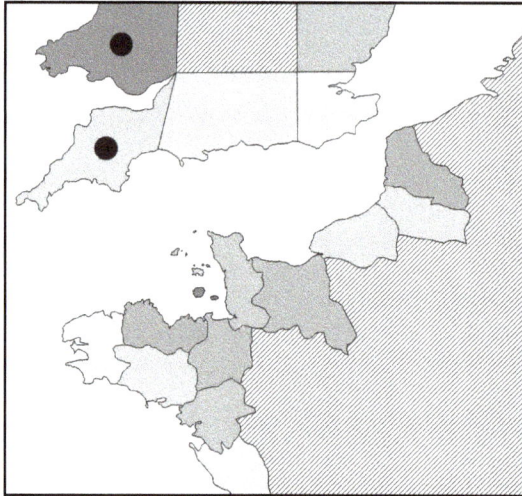

This colonial bryozoan is native to North-west Europe has been established on the west coast of the USA for some time. In 2013 it was found in Welsh ports and has since spread to locations in Scotland. It is thought that this species has the ability to spread rapidly between ports via leisure shipping. Reports have so far been confined to marinas and harbours where it can occur in extensive sheets. It is felt that there is the potential for *S. japonica* to spread into the wider marine environment. Although some distance from the Channel Islands, the rate of spread and experience from other parts of the world suggests that it may quickly colonise harbours within the English Channel. Regular checks should be made for this species.

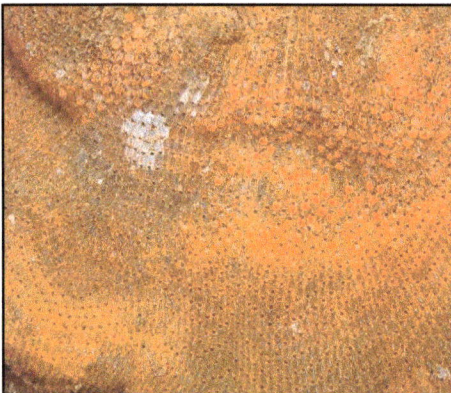

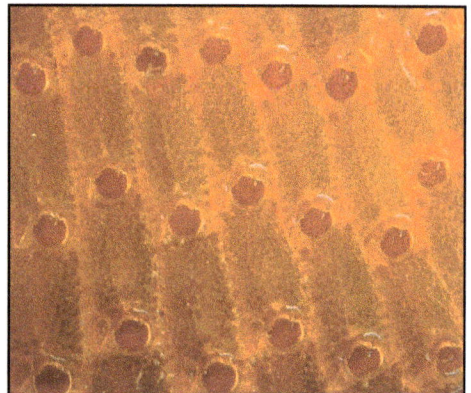

# *Victorella pavida*
## Bryozoan

Threat score: 1

| ES | Hab | Tox | Econ |
|----|-----|-----|------|
| 1 | 1 | 1 | 1 |

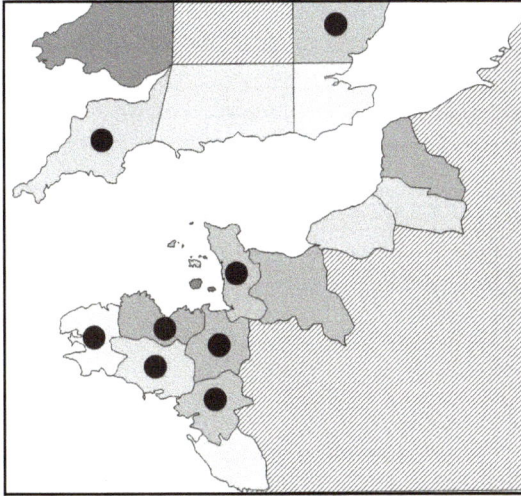

A cryptogenic species that was first identified in London in the 1860s but which was not recorded from Brittany until the 1960s. Primarily a species of fresh and brackish waters but there are references to it tolerating higher salinities.

It has been suggested that the optimum salinity for this species is around 10 to 18 ppt and as such it is unlikely ever to be present in the fully marine water surrounding the Channel Islands.

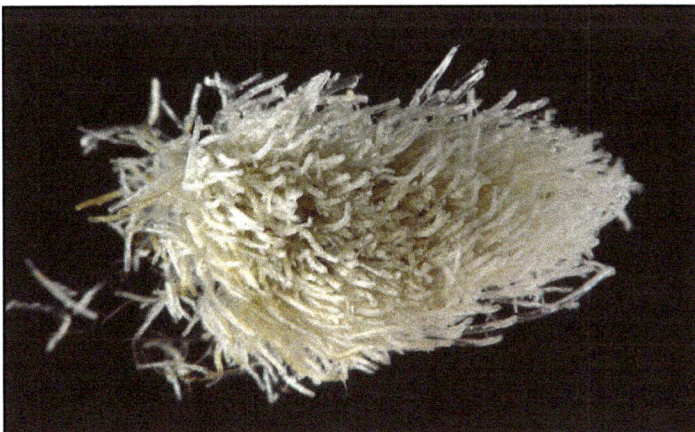

# Didemnum vexillum
## Carpet Sea Squirt

Threat score: 80

| ES | Hab | Tox | Econ |
|----|-----|-----|------|
| 4 | 4 | 1 | 5 |

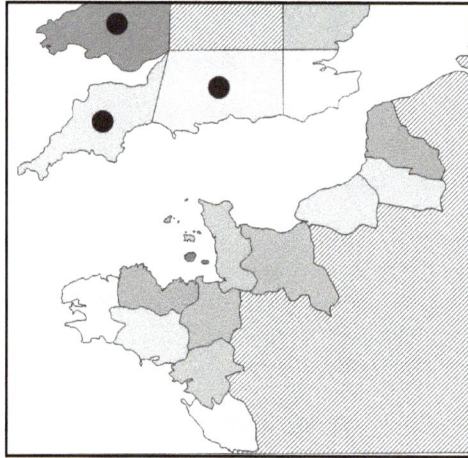

A colonial ascidian from the North-west Pacific *D. vexillum* may have been introduced into Europe as early as the 1970s via aquaculture imports. The species was officially recognised in 1998 from northern France and during the past decade there have been reports from Ireland (2006), UK (2008) and English Channel ports.

Primarily a species of harbours and marinas, the ability of *D. vexillum* to overwhelm native species and habitats is a major cause of concern and it is widely considered to be a serious threat to habitats and species as well as being a potentially serious fouling pest. *D. vexillum* is spreading rapidly with shipping believed to be the main transfer mechanism and recent records from the Solent and Isle of Wight suggest that it will be present in the Channel Islands in the near future. Searches of the marinas at St Helier have so far revealed no sign of *D. vexillum* but is should be considered as a potential threat to the Channel Islands.

*Left:* D. vexillum *growing in an aquaculture area in the Netherlands.*
*Right: Detail of a colony of* D. vexillum.

136

# *Botrylloides violaceus*
## Sea Squirt

Threat score: 8

| ES | Hab | Tox | Econ |
|----|-----|-----|------|
| 2 | 4 | 1 | 1 |

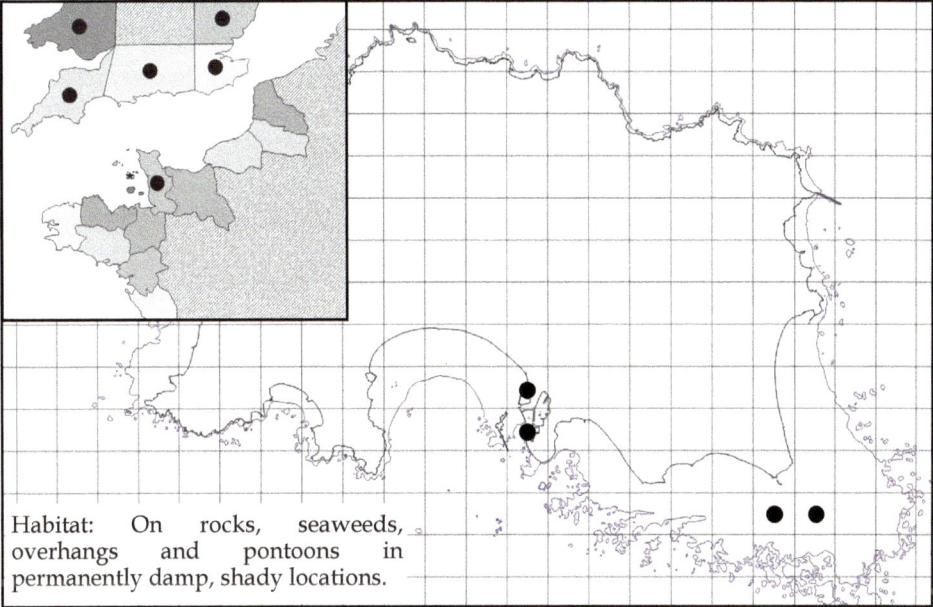

Habitat: On rocks, seaweeds, overhangs and pontoons in permanently damp, shady locations.

Introduced into Europe via aquaculture during the 1990s, *Botrylloides violaceus* has spread rapidly, probably via shipping, to marinas and harbours along the Atlantic and English Channel coasts. In Jersey it was first identified at La Rocque in 2016 but its similarity to a local species (*Botrylloides leachii*) makes it probable that it has been here for several years. For example, photographs taken of a specimen identified as *B. leachii* in 2009 are probably of *B. violaceus*. It is uncommon on the seashore and has shown little sign of spreading rapidly. It undoubtedly competes with local under-boulder fauna but is at present not widespread or abundant enough to present a serious threat.

*Left:* Botrylloides violaceus *under a rock at La Rocque Harbour in 2016.*
*Right: Zooids from the same specimen viewed through a microscope.*

137

# *Botrylloides diegensis*
## Sea Squirt

Threat score: 8

ES   Hab   Tox   Econ
2    4     1     1

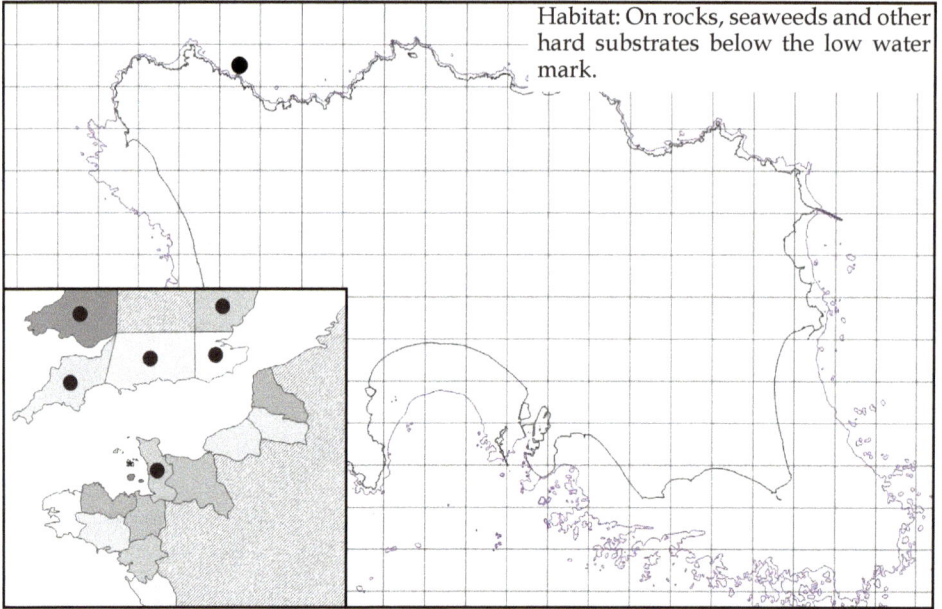

Habitat: On rocks, seaweeds and other hard substrates below the low water mark.

First found in the Netherlands in 2002, *B. diagensis* has since spread to southern England, Wales and Normandy where it is frequently associated with harbours and marinas.

The first Channel Islands records were made a matter of days before this report was due to be finalised when a specimen on a Spider Crab caught off Guernsey was identified by Richard Lord. Two weeks later a subtidal specimen was identified off the north coast of Jersey by Dr Lin Baldock. It is probable that this species is a recent arrival but nonetheless may already be widespread within the islands. It has not been identified in any island harbours or marinas suggesting it may have spread here naturally, perhaps from the Normandy coast.

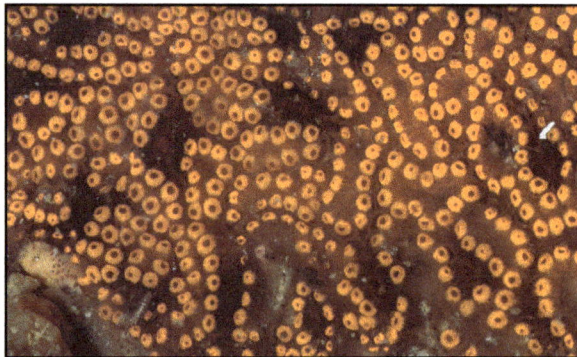

*Botrylloides diagensis on a spider crab caught off Guernsey in August 2017.*

# *Perophora japonica*
## Sea Squirt

Threat score: 18

| ES | Hab | Tox | Econ |
|----|-----|-----|------|
| 3 | 3 | 1 | 2 |

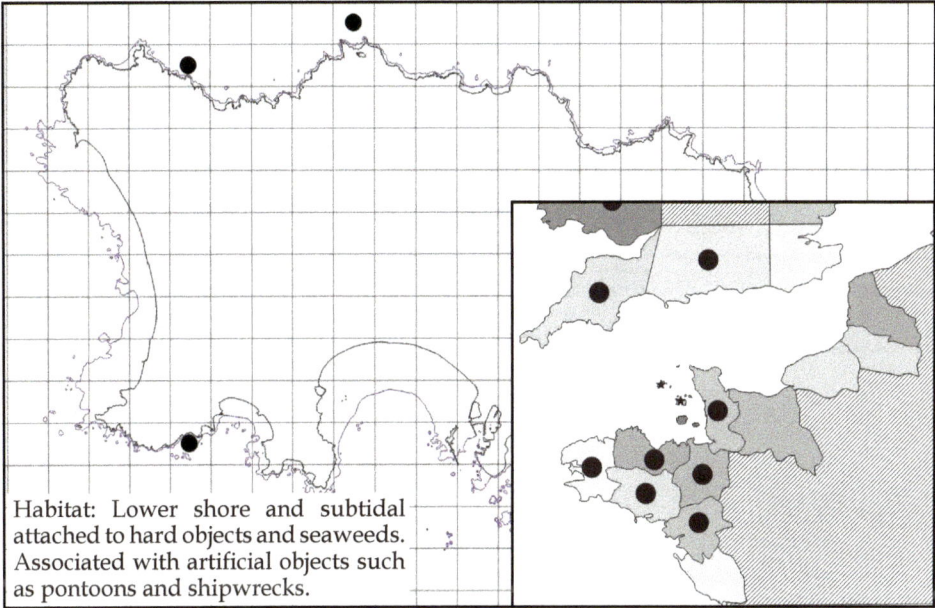

Habitat: Lower shore and subtidal attached to hard objects and seaweeds. Associated with artificial objects such as pontoons and shipwrecks.

First recorded in northern France in the early 1980s, by 1999 it was in Plymouth. It was not identified in the Channel Islands until 2003 but was probably established some years previously. *P. japonica* was found offshore in Jersey in 2013 on a shipwreck where an estimated third of the wreck's surface was covered by it.

*P. japonica* requires specialist knowledge to identify it which, combined with its subtidal habitat, means that it is probably much more widespread and common in the Channel Islands than the handful of records would suggest. It has the ability both to foul structures and displace native species but its actual status, and therefore threat level, is otherwise difficult to determine at present.

Perophora japonica *photographed in Guernsey in 2003. The distinctive yellow star-shaped stolon terminal buds are visible in the left image.*

# *Corella eumyota*
## Sea Squirt

Threat score: 27

| ES | Hab | Tox | Econ |
|----|-----|-----|------|
| 3 | 3 | 1 | 3 |

Habitat: Lower shore and subtidal attached to rocks. Associated with artificial objects such as pontoons and shipwrecks.

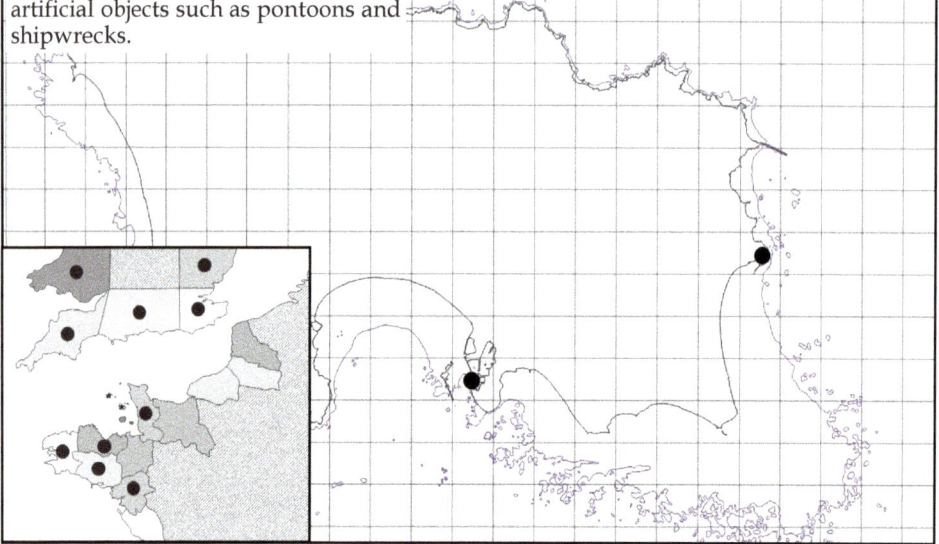

Originally from the South Pacific, this solitary sea squirt was first recorded in Brittany and the UK in 2004 and has since spread to many other locations. Although primarily known from marinas, *C. eumyota* has recently been found on seashores adjacent to ports in southern England. It is potentially a fouling threat and if established in the wild may compete for space with native encrusting animals.

Known from Guernsey since 2007, it was recently identified in Jersey both at St Helier harbour and on the seashore.

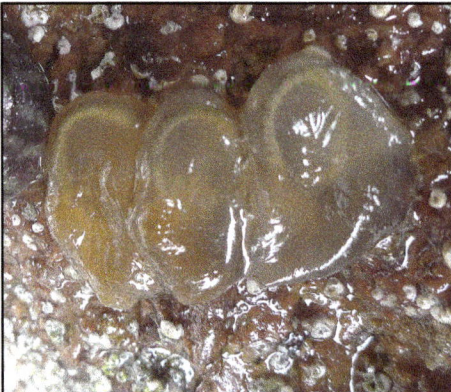

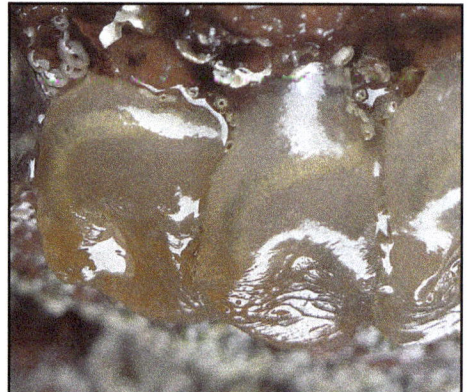

*Left: Specimens of* **Corella eumyota** *on rocks behind Gorey Pier, Jersey. Right: The same specimens showing the characteristic u-shaped intestine.*

140

# *Molgula manhattensis*
## Sea Grape

Threat score: 18

| ES | Hab | Tox | Econ |
|----|-----|-----|------|
| 3 | 3 | 1 | 2 |

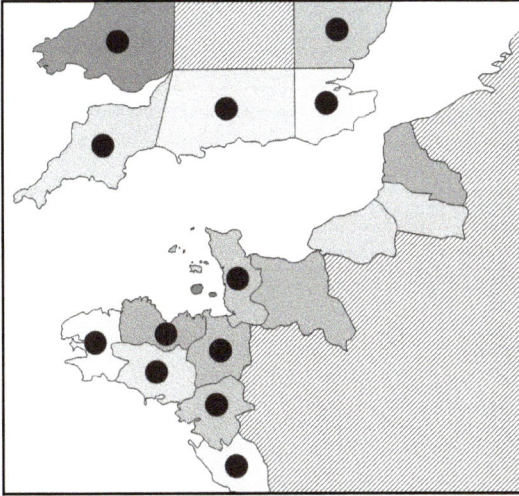

A small solitary tunicate that is native to the eastern coast of North America. It may have been in Europe as early as the 1840s and is now widespread; it has been recorded on all parts of the Normandy and Brittany coast. Noted as a fouling pest, *M. manhattensis* may form dense colonies, especially in ports and harbours, but can be found on hard substrates from the lower shore to depths of 90 metres. Although not formally identified from the Channel Islands, it may have been established in these waters for some time but perhaps does not occur in dense colonies.

# Styela clava
## Leathery Sea Squirt

Threat score: 27

| ES | Hab | Tox | Econ |
|----|-----|-----|------|
| 3 | 3 | 1 | 3 |

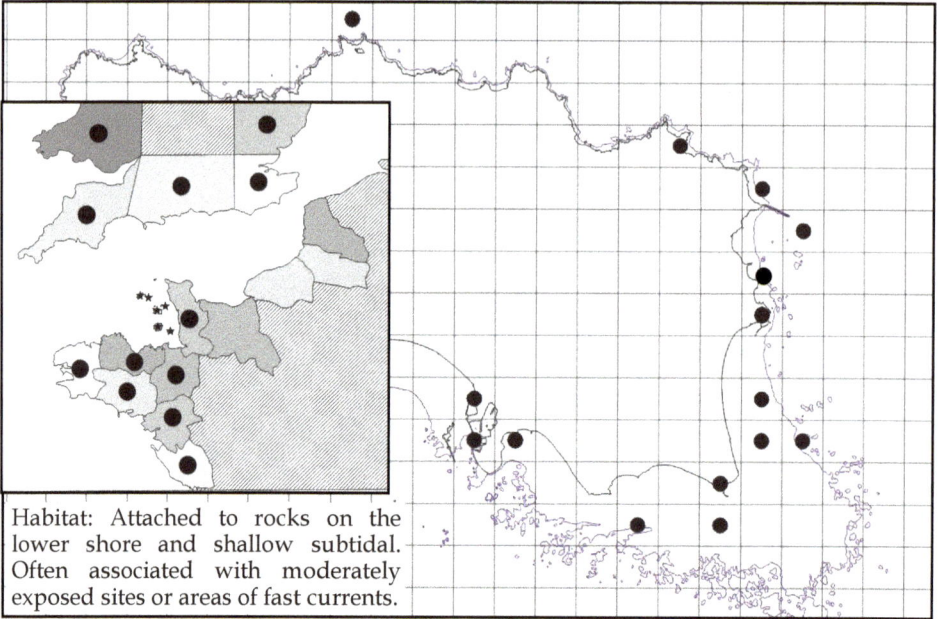

Habitat: Attached to rocks on the lower shore and shallow subtidal. Often associated with moderately exposed sites or areas of fast currents.

Originally identified in Plymouth in 1953, it is probable that *Styela clava* arrived on warships returning from the Korean War. Its subsequent spread in the UK and then Europe was rapid and may have been facilitated by both the movement of shipping and aquaculture stock. It reached northern France in 1968 but was not recorded in the Channel Islands until 1998 although it was probably established much earlier than this.

Although common in the marinas at St Peter Port and St Helier, *S. clava* also occurs frequently in the subtidal, especially in more exposed rocky locations such the offshore reefs. It is at relatively low densities on the seashore and offshore which, with its compact holdfast and isolated occurrence, makes it a moderate to low level threat to local species. Its population seem stabilised in the subtidal areas but it might still be spreading intertidally. This is a species that is worth monitoring.

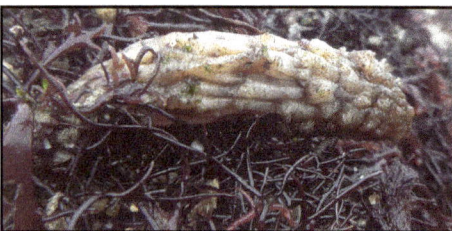

*Specimens photographed in Jersey waters.*

# *Asterocarpa humilis*
## Waxy Sea Squirt

Threat score: 27

| ES | Hab | Tox | Econ |
|----|-----|-----|------|
| 3  | 3   | 1   | 3    |

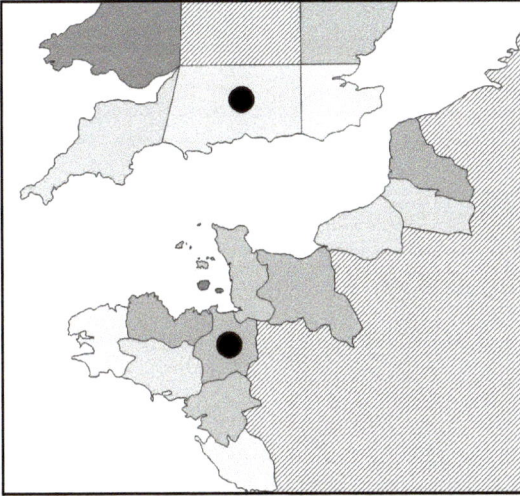

Native to the southern pacific Ocean, *A. humilis* has been transported to several parts of the world and in Europe was first recorded in 2005 outside St Malo in northern Brittany. By 2010 it had reached southern England and is believed to still be spreading along the English Channel.

There are no Channel Island records but given its occurrence elsewhere in the Normano-Breton Gulf it seems probable that *A. humilis* may already be present or will be in the very near future. It is primarily a species of ports and marinas where, like other ascidians, it can cause problems with fouling.

# *Oncorhynchus kisutch*
## Coho Salmon

Threat score: 2

| ES | Hab | Tox | Econ |
|----|-----|-----|------|
| 1 | 2 | 1 | 1 |

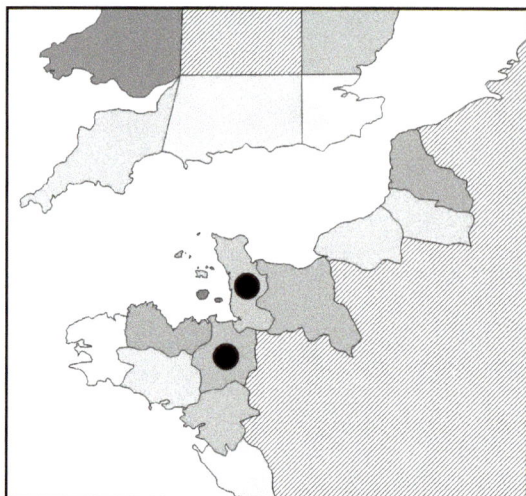

There are four non-native species of *Oncorhynchus* (fish from the salmon family) with reports in the British Isles, two of which have regional reports. In 1977 a single specimen of *O. kisutch* (Coho Salmon) was caught off St Sampson in Guernsey and in 1984 a single specimen of *O. mykiss* (Rainbow Trout) was caught at the Rance Barrage in Brittany. Both species are native to the NE Pacific and both are associated with aquaculture in Europe.

Since the nineteenth century *O. mykiss* has been introduced into a large number of reservoirs and lakes as a sport fish (including in Jersey) and is extremely widespread. *O. kisutch* has been selectively reared in aquaculture facilities in northern France and between 1973 and 1974 at least 50,000 individuals escaped into the wild in Normandy alone. This is thought to be responsible for anglers having caught 25 adult fish in rivers and coastal regions around Normandy between 1975 and 1977. This is the probable origin of the lone Guernsey specimen and a lack of further records suggests that *O. kisutch* is not established locally.

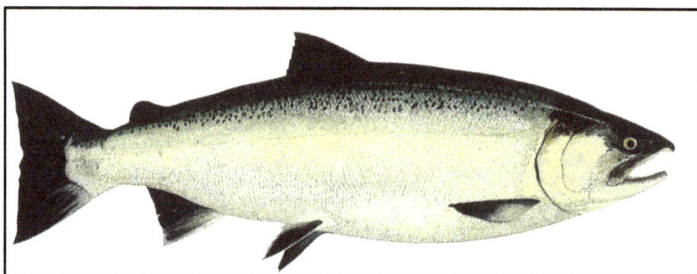

*A drawing of the male ocean phase Coho Salmon.*

144

# 5 - Non-native Marine Species
## *- Part Two: Plants -*

Gomez (2008) argues that many non-native phytoplankton species in the English Channel have been historically misidentified. This may mean that some phytoplankton species that are currently listed as non-native (by some authorities at least), may in fact have a much longer regional track record than the literature suggests or even be cosmopolitan species. We have included these controversial species in this report but their status may need to be revised at a later date.

# *Stephanopyxis palmeriana*
## Diatom

Threat score: 1

ES  Hab  Tox  Econ

| 1 | 1 | 1 | 1 |

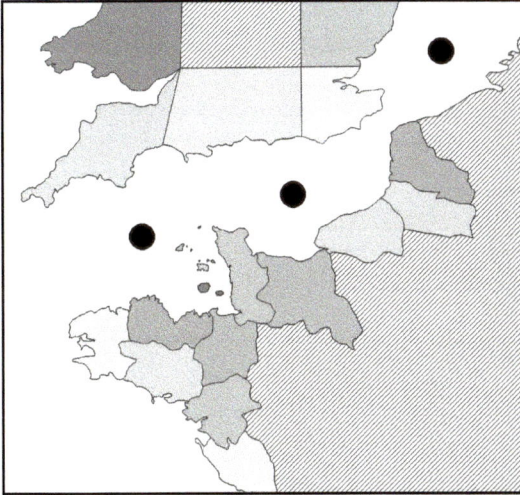

A Pacific diatom which has been recorded from the North Sea and eastern English Channel since at least the 1950s. It is probably present in Channel Island waters but is not thought to pose a serious threat. Considered not to be a non-native species by Gomez (2008).

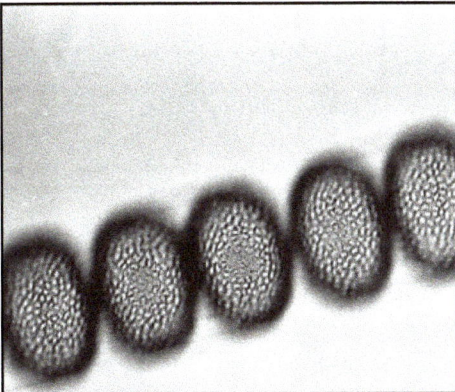

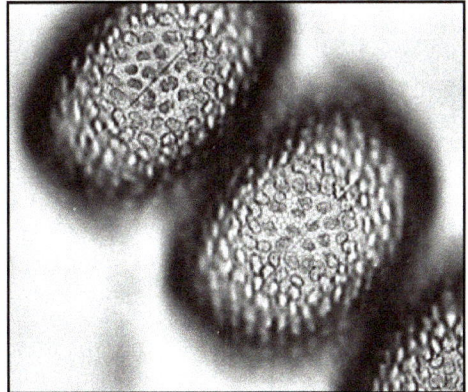

# *Thalassiosira punctigera*
## Diatom

Threat score: 1

| ES | Hab | Tox | Econ |
|----|-----|-----|------|
| 1 | 1 | 1 | 1 |

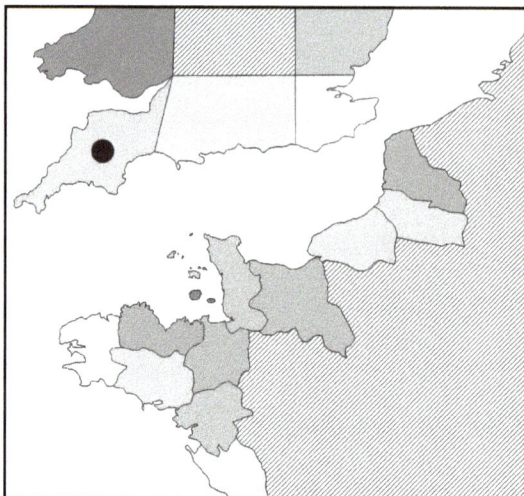

Originally from the Pacific but it became common in the North Atlantic during the 1970s and is now cosmopolitan. Known from the English Channel and probably present in Channel Island waters but it is not thought to pose a serious threat. Considered not to be a non-native species by Gomez (2008).

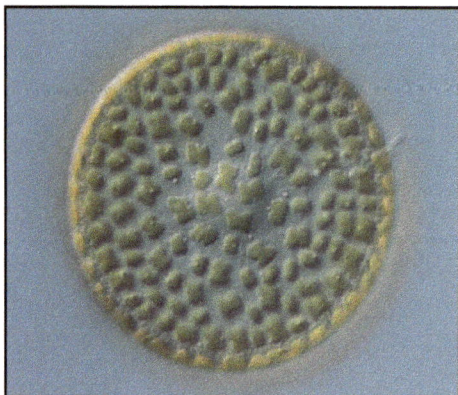

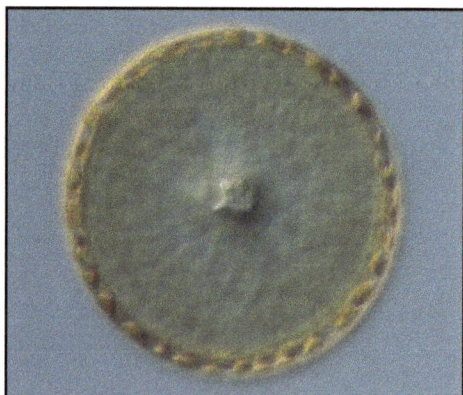

# *Thalassiosira tealata*
## Diatom

Threat score: 2

| ES | Hab | Tox | Econ |
|----|-----|-----|------|
| 2 | 1 | 1 | 1 |

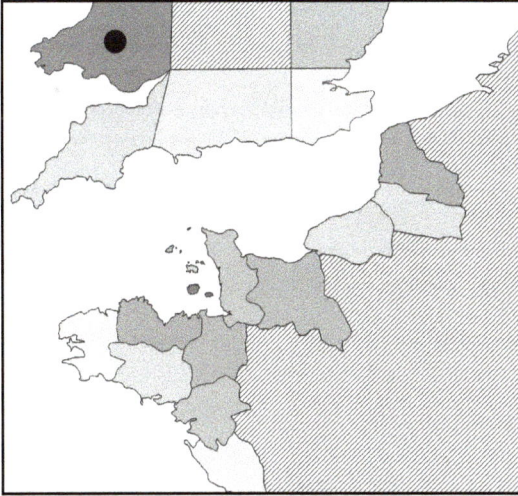

Native to the Pacific Ocean but known from northern Europe (including the English Channel) since at least the 1990s. Possibly present in Channel Island waters and is not thought to pose a serious threat. Considered not to be a non-native species by Gomez (2008).

# *Coscinodiscus wailesii*
## Diatom

Threat score: 12

| ES | Hab | Tox | Econ |
|----|-----|-----|------|
| 2 | 3 | 1 | 2 |

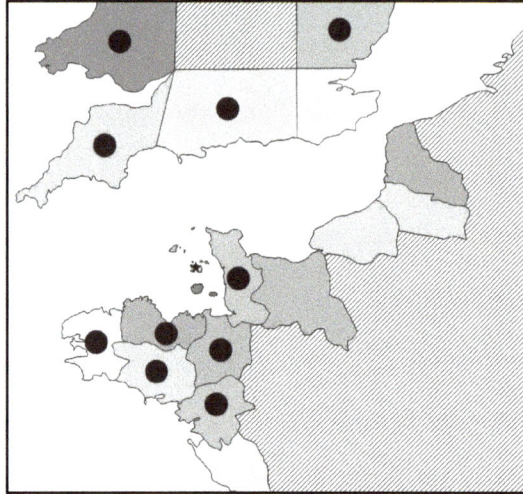

Introduced into European waters in the late 1970s from the Pacific and now widely distributed in the English Channel. A large diatom that can produce large quantities of mucilage (slime) and which is subject to algal blooms in the winter and spring. Blooms have been associated with mucilage clogging fishing nets and there are concerns that the size of this diatom and its dominance during blooms may affect the feeding ability of smaller organisms such as copepods, which cannot digest it.

Confirmed from Jersey in October 2016, *C. wailesii* is probably common in local waters. Fractured valves of what was probably *C. wailesii* were found in a spume (sea foam) event in St Aubin's Bay during the summer of 2015. In some countries *C. wailesii* has been recommended for long-term observation with regard to its effect on the environment and in particular the food chain.

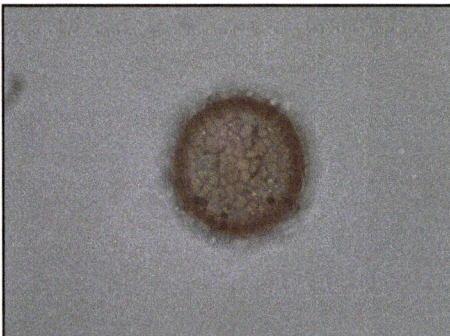

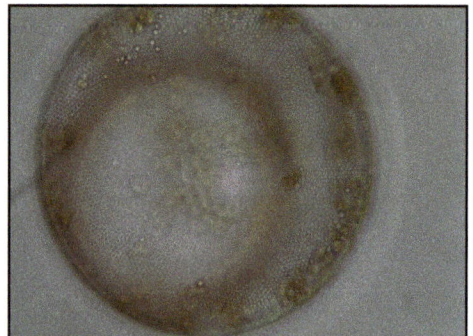

*Above: A specimen of* C. wailesii *from a plankton sample taken off Jersey in 2016.*

# *Odontella sinensis*
## Diatom

Threat score: 2

| ES | Hab | Tox | Econ |
|----|-----|-----|------|
| 1  | 2   | 1   | 1    |

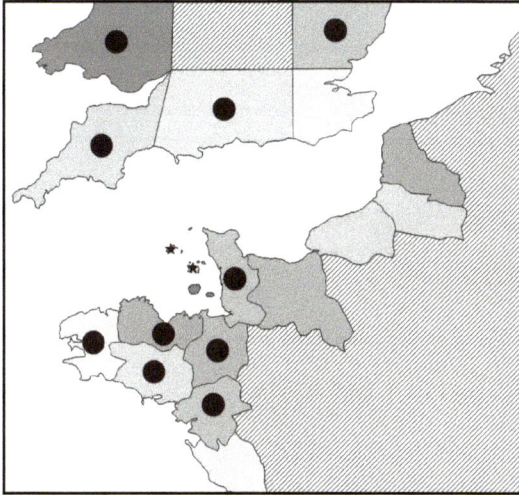

Originally from the Pacific Ocean, European specimens of *O. sinensis* were first observed in the early twentieth century. Considered to be cosmopolitan (and possibly even cryptogenic in nature rather than non-native) in its distribution and to pose little threat to local species of habitats. *O. sinensis* has been recorded from both Jersey and Guernsey waters. Considered not to be a non-native species by Gomez (2008).

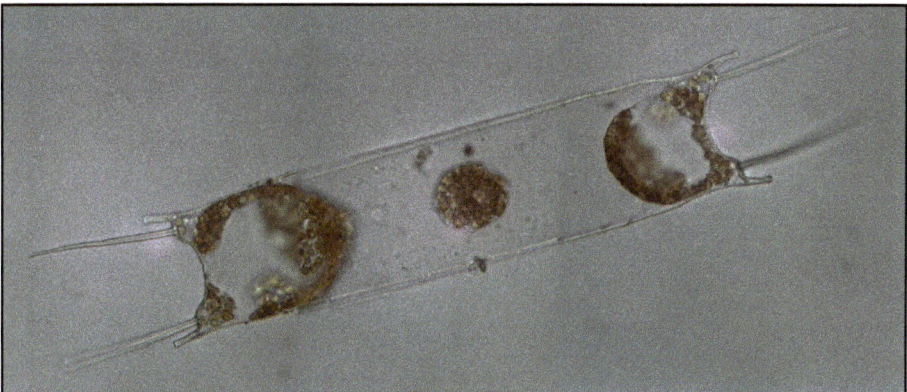

*Above: A specimen of* O. sinensis *from a plankton sample taken off Jersey's east coast.*

# *Corethron pennatum*
## Diatom

Threat score: 1

| ES | Hab | Tox | Econ |
|----|-----|-----|------|
| 1 | 1 | 1 | 1 |

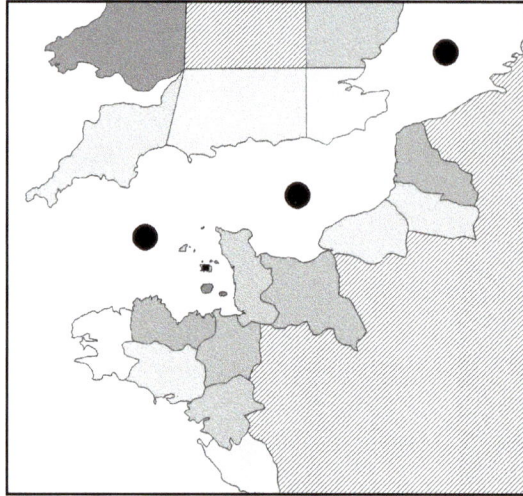

Native to the Pacific but now widespread around the globe including in the English Channel where it was first recorded in the 1950s. Specimens were found off Jersey in the autumn of 2016 but it is not thought to be common and is not thought to pose a serious threat to the lcoal environment. Considered not to be a non-native species by Gomez (2008).

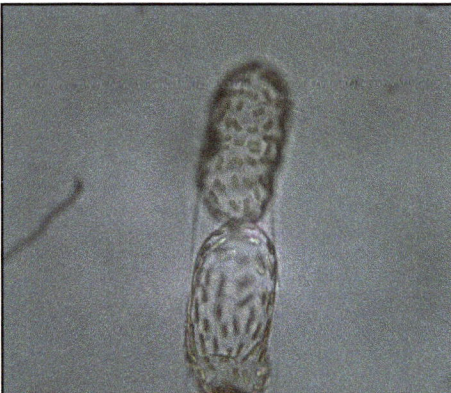

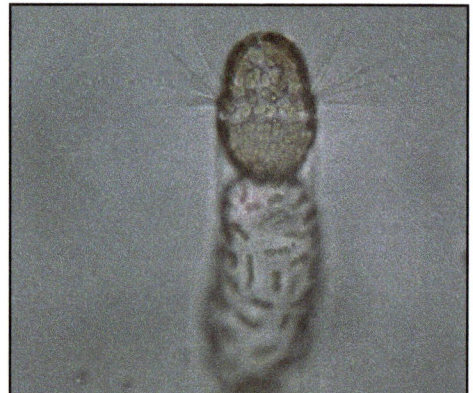

*Above:* Corethron pennatum *from a plankton sample taken off Jersey's south coast.*

# *Pleurosigma simonsenii*
## Diatom

Threat score: 1

| ES | Hab | Tox | Econ |
|----|-----|-----|------|
| 1 | 1 | 1 | 1 |

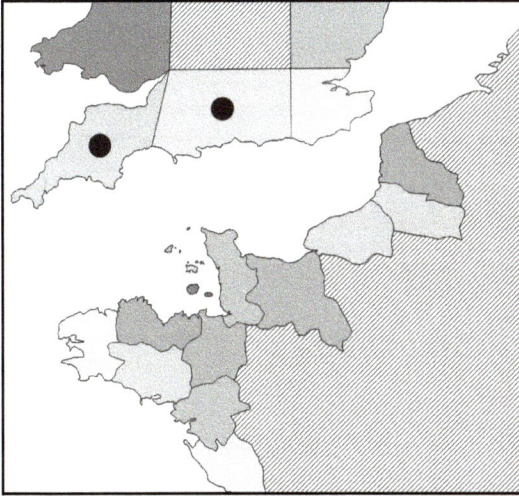

Known from the English Channel since at least the 1990s. Not thought to pose a serious threat and is possibly present in Channel Island waters. Considered not to be a non-native species by Gomez (2008).

# *Pseudo-nitzschia multistriata*
## Diatom

Threat score: 1

| ES | Hab | Tox | Econ |
|----|-----|-----|------|
| 1 | 1 | 1 | 1 |

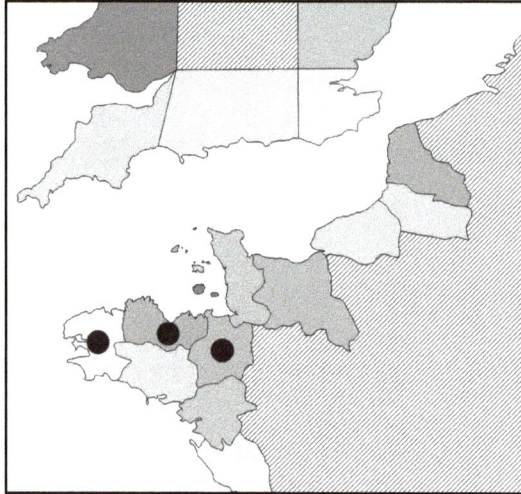

Recorded from the northern Brittany coast. Species of *Pseudonitzschia* are known from Jersey waters and are associated with ASP toxins but whether *P. multistriata* is one of these is not known. It is probably present in Channel Island waters.

# *Alexandrium affine*
## Dinoflagellate

Threat score: 9

| ES | Hab | Tox | Econ |
|----|-----|-----|------|
| 1  | 1   | 3   | 3    |

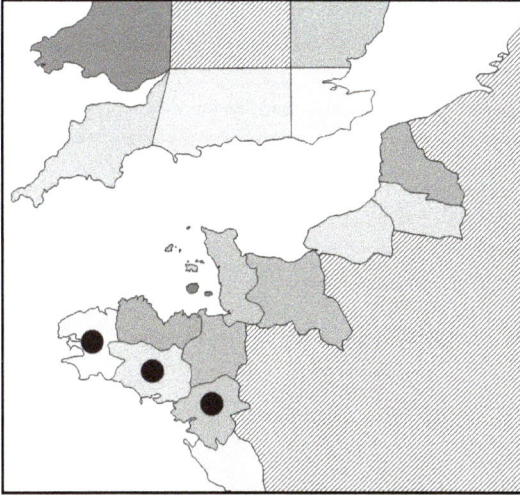

Native to the Pacific Ocean but recorded from the western Atlantic including northern and southern Brittany in 1987. It's status in Channel Island waters is unknown.

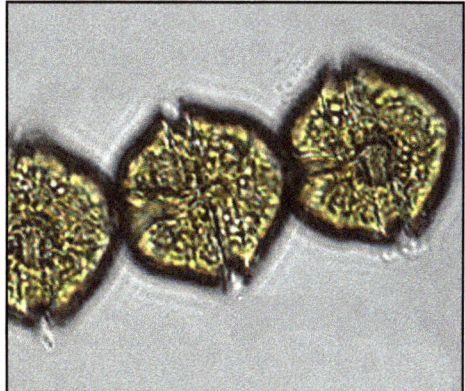

# *Alexandrium leei*
## Dinoflagellate

Threat score: 9

| ES | Hab | Tox | Econ |
|----|-----|-----|------|
| 1 | 1 | 3 | 3 |

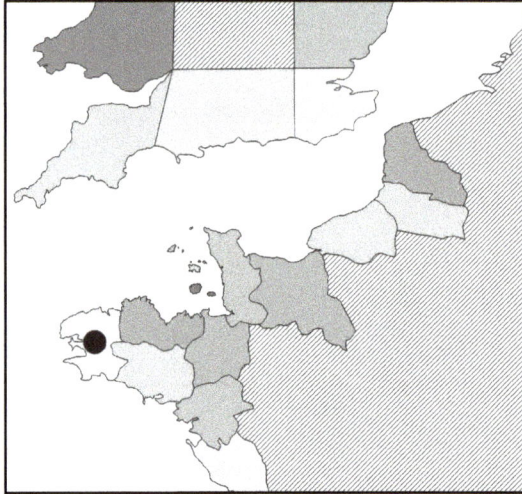

A North-west Pacific species recorded in northern Brittany in 1993. Noted in AlgaeBase as being harmful to other species but with no further details. *A. leei* may be present in Channel Island waters. Considered not to be a non-native species by Gomez (2008).

# *Alexandrium minutum*
## Dinoflagellate

Threat score: 24

| ES | Hab | Tox | Econ |
|----|-----|-----|------|
| 2 | 2 | 3 | 2 |

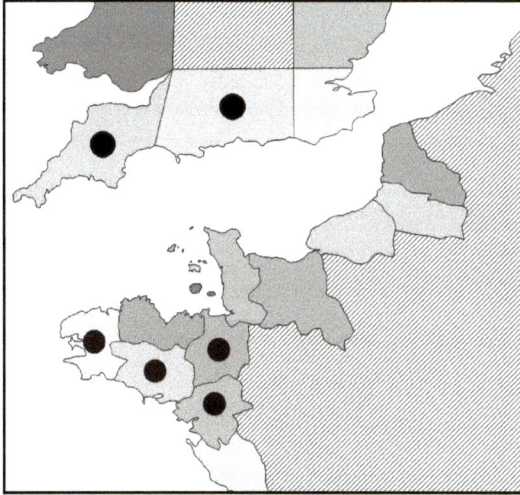

Associated with toxic paralytic shellfish poisoning events that can affect mammals (including humans), birds and fish. Toxicity events have occurred in Brittany (notably near Brest in 2012) but these seem to be triggered within estuaries by the combination of a high nutrient river discharges in conjunction with high tides. Although *A. minutum* may be present locally it is unlikely to present a threat to Jersey's coast. Considered not to be a non-native species by Gomez (2008).

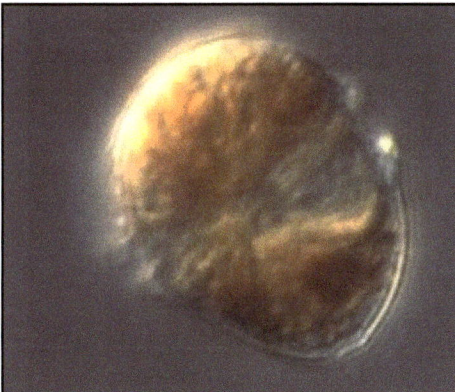

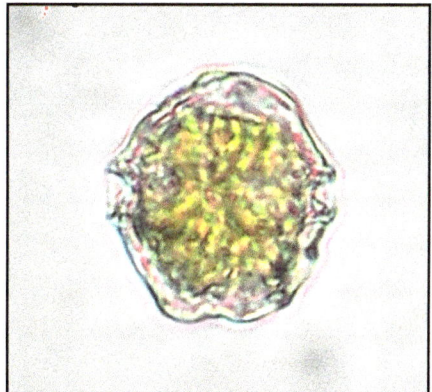

# *Karenia brevisulcata*
## Dinoflagellate

Threat score: 9

| ES | Hab | Tox | Econ |
|----|-----|-----|------|
| 1  | 1   | 3   | 3    |

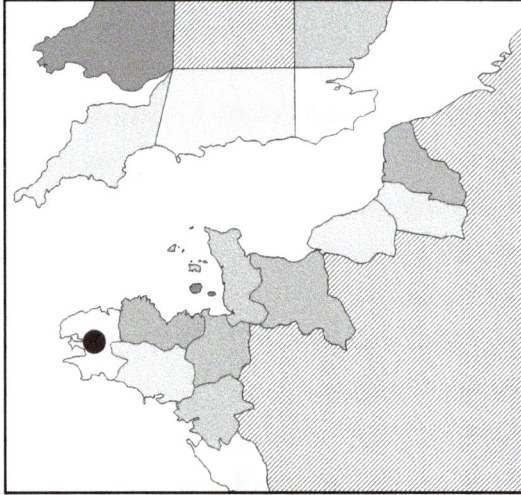

A North-west Pacific species first recorded in northern Brittany in 1970. Algal blooms of this species have been associated with harmful events in Wellington Harbour, New Zealand. Its status within the Normano-Breton Gulf is unknown but as a planktonic marine algae, *K. brevisulcata* may be present in Channel Island waters.

# *Karenia papilionacea*
## Dinoflagellate

Threat score: 8

| ES | Hab | Tox | Econ |
|----|-----|-----|------|
| 1  | 1   | 4   | 2    |

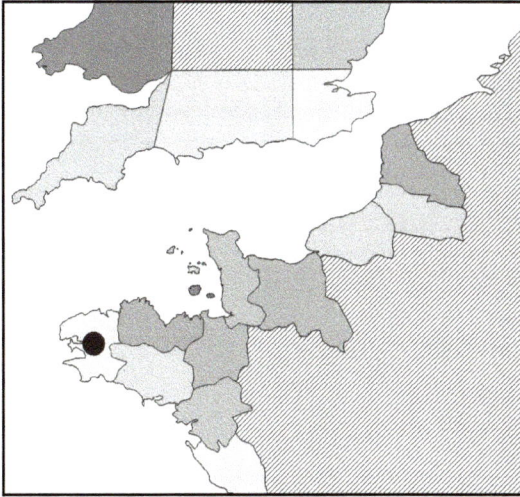

Native to the North-west Pacific and first recorded in Brittany in 1994. This species can be associated with paralytic poisoning events within sea urchins, molluscs and fish. As with *Alexandrium minutum* toxicity events seem to be mainly associated with estuarine conditions and are unlikely to occur in Jersey.

# *Karenia umbella*
## Dinoflagellate

Threat score: 8

| ES | Hab | Tox | Econ |
|----|-----|-----|------|
| 1  | 1   | 4   | 2    |

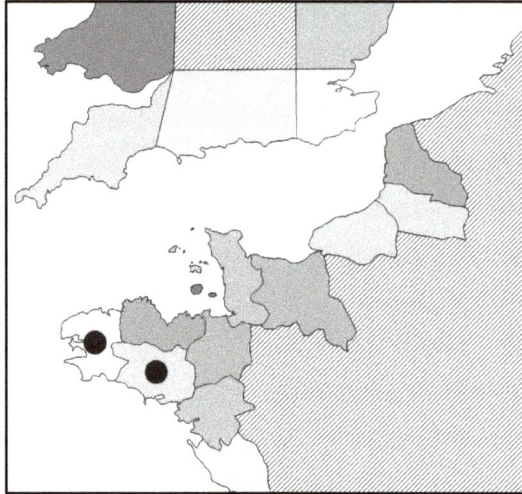

A South-west Pacific species recorded from southern Brittany in 2008. Noted as a species that is potentially toxic to fish. It has not been recorded from the Normano-Breton Gulf but may be present in low abundance.

# *Takayama tasmanica*
## Dinoflagellate

Threat score: 8

| ES | Hab | Tox | Econ |
|----|-----|-----|------|
| 1  | 1   | 4   | 2    |

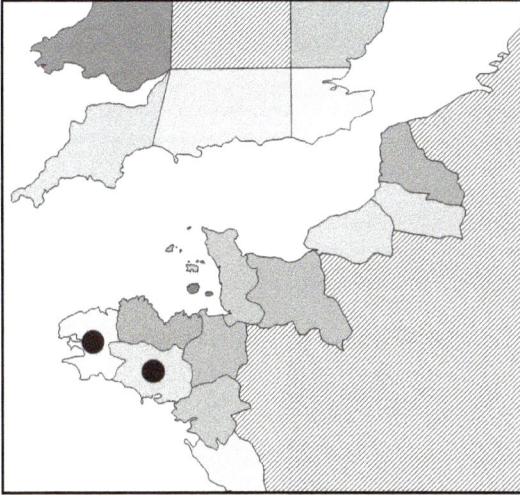

A South-west Pacific species that was reported from southern Brittany in 2008 but which has since been reported from north Brittany also. It has been associated with fish kill events in Florida estuaries but is unlikely to present a threat in local waters.

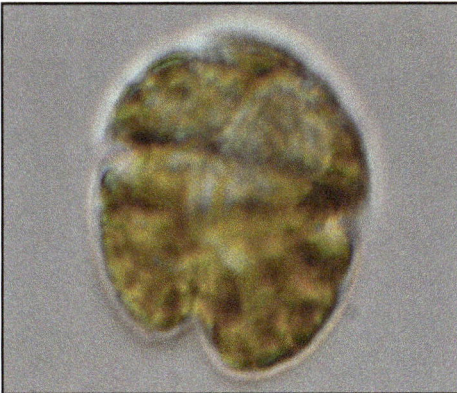

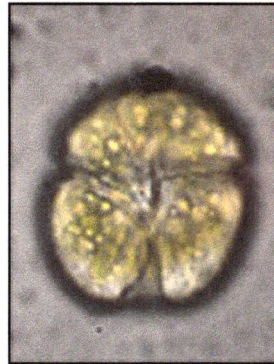

# *Fibrocapsa japonica*
## Ochrophyta

Threat score: 4

| ES | Hab | Tox | Econ |
|----|-----|-----|------|
| 1  | 1   | 2   | 2    |

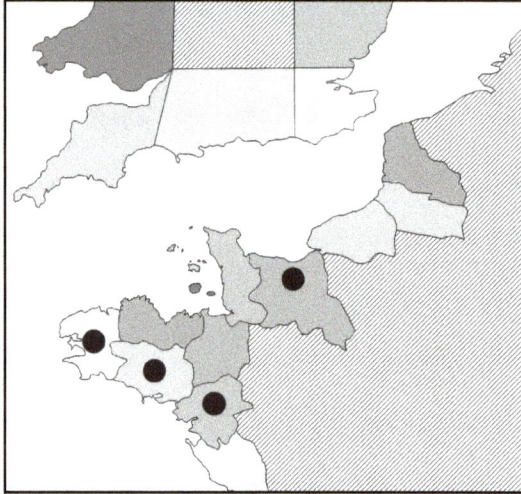

A marine algae from the North-west Pacific which has been associated with mass fish mortality events in Japan. It has been known from European waters since the early 1990s and has been widely reported including from the Normandy and Brittany coasts. *F. japonica* has not been associated with fish deaths in Europe and it is currently not regarded as a threat. The widespread and abundant nature of this species suggests that it is probably in Channel Island waters but is unlikely to present a threat.

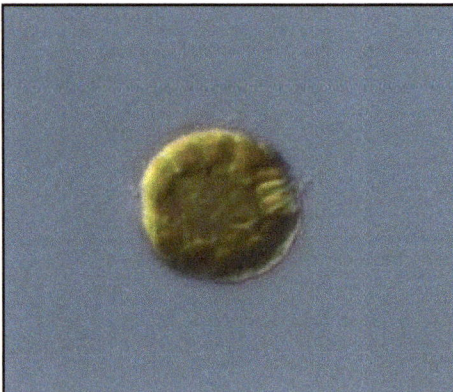

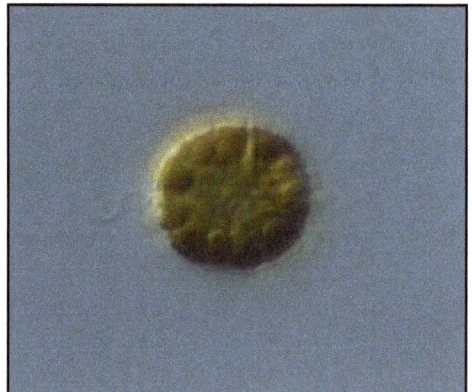

# Heterosigma akashiwo
## Ochrophyta

Threat score: 48

| ES | Hab | Tox | Econ |
|----|-----|-----|------|
| 2 | 2 | 4 | 3 |

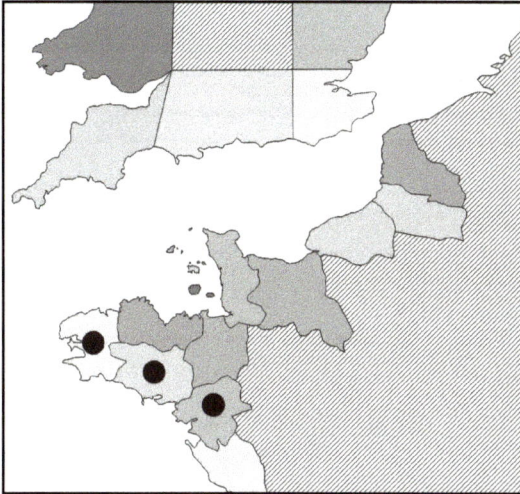

A coastal microscopic algae that has been associated with highly toxic red tides that have killed thousands of farmed salmon in British Colombia. *H. akashiwo* is native to the North-west Pacific Ocean but has become widespread in temperate and tropical waters across the globe. Although reported from the Normano-Breton Gulf, there have been no toxic events (which seem to be associated with estuaries) associated with this species in the region. As a primarily coastal species, *H. akashiwo* may well be present in Channel Island waters but it is unlikely to form red tide events.

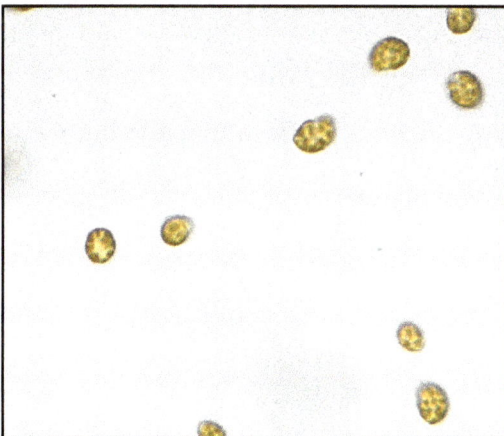

# *Undaria pinnatifida*
## Wakame

Threat score: 36

| ES | Hab | Tox | Econ |
|----|-----|-----|------|
| 3 | 4 | 1 | 3 |

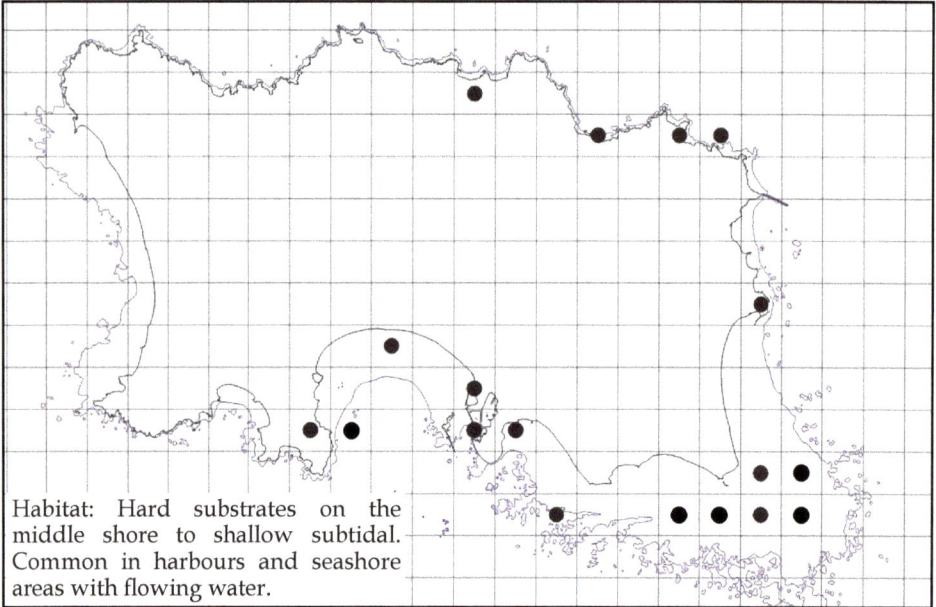

Habitat: Hard substrates on the middle shore to shallow subtidal. Common in harbours and seashore areas with flowing water.

First introduced into the Mediterranean in the 1970s, *Undaria pinnatifida* has since been spread via deliberate aquaculture and shipping to many parts of Europe. It was recorded on the north Brittany coast in 1984 and was probably in the Channel Islands within a decade although it was not officially identified until some years later.

For many years *U. pinnatifida* seemed to be restricted to the harbours in Guernsey and Jersey but by 2009 isolated populations had been observed in intertidal areas on both islands. Once established at a location, *U. pinnatifida* returns annually in the late winter and spring but often dies back in the early to mid-summer.

In Jersey the number of known seashore locations has increased steadily since 2009 but during the spring of 2016 the distribution and abundance of *U. pinnatifida* increased markedly particularly along the south and south-east coasts. The cause of this sudden spread is unknown but Wakame has become abundant in some locations and is even competeing with Wireweed (*Sargassum muticum*) for space. It appears to be absent from the west coast and, as with *Watersipora subatra*, may be colonising the seashore by spreading east and west of St Helier.

*U. pinnatifida* has the potential to displace native seaweed species including *Sargassum muticum*, another proflific non-native species. As an actively spreading large seaweed it should be considered as a potential ecological and fouling threat. Close monitoring of populations on the seashore and in the subtidal is recommended.

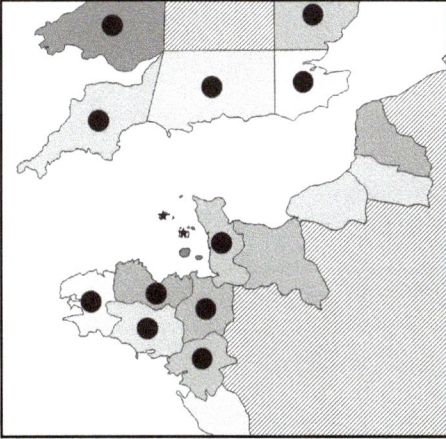

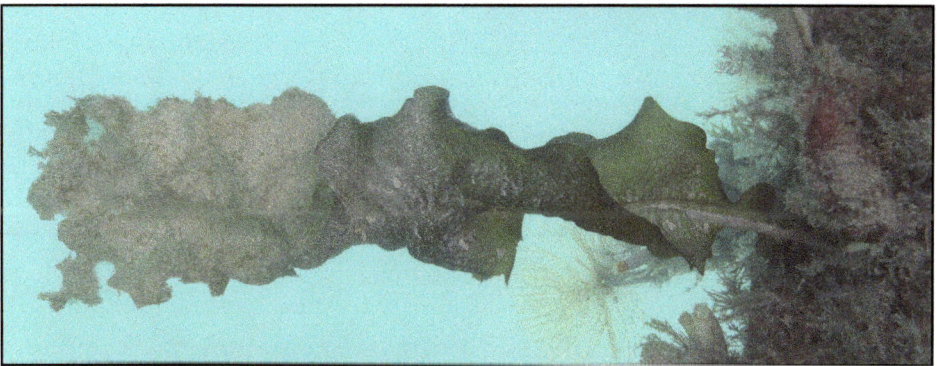

*Top right: Wakame competing for space on Jersey's south coast with another non-native species of seaweed, **Sargassum muticum**. Middle: A mature specimen of Wakame growing on a pontoon in St Helier Marina. Top right: A young specimen growing in the early spring at La Collette Marina, Jersey. Note the seaweed's distinctive fingered shape along the length of the stipe. Bottom left: A mature specimen growing on the shore at Green Island, Jersey, showing the ridged patterning along the base of its stipe. By late summer this ridging may be all that's left of plants growing on the seashore.*

# Sargassum muticum
## Wireweed

Threat score: 100

| ES | Hab | Tox | Econ |
|----|-----|-----|------|
| 5 | 5 | 1 | 4 |

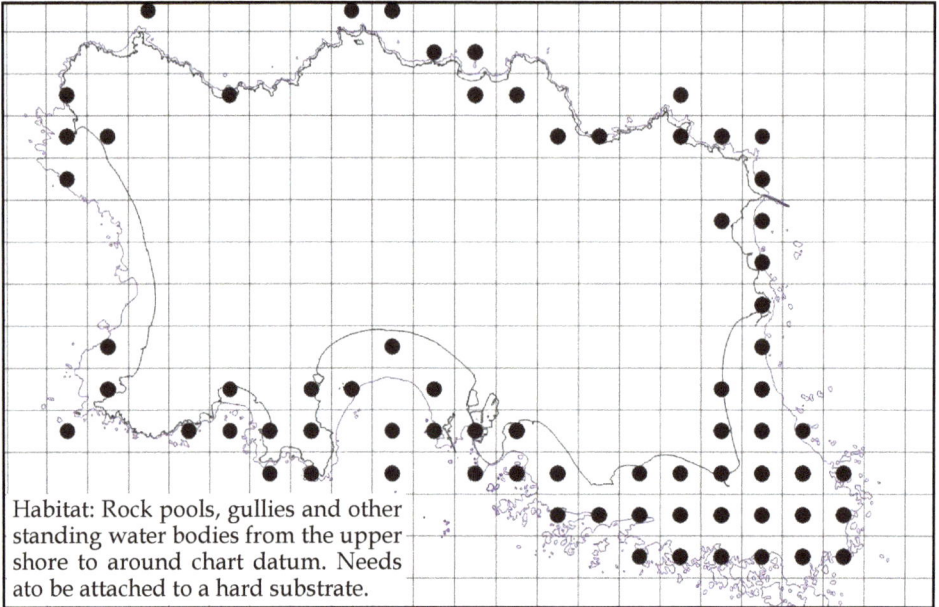

Habitat: Rock pools, gullies and other standing water bodies from the upper shore to around chart datum. Needs ato be attached to a hard substrate.

Wireweed has probably had the greatest visual impact of any non-native marine species within the Channel Islands. The species was probably introduced into Europe in the mid-1960s via aquaculture stock imported from the North-west Pacific. In the late 1970s it had spread rapidly within the English Channel, reaching the Channel Islands in 1979. Attempts were made to iradicate Wireweed from the seashore but the species established itself and had become prolific by the early 1980s.

Some specimens of Wireweed can grow to eight metres in length which, combined with its bushiness and ability to form dense strands, can cause it to dominate rock pools, shallow marine areas and intertidal bodies of water. The effects have been particularly noticable in the flooded gully complexes on Jersey's east coast and the shallow marine areas of all the Channel Islands.

Given its size and ubiquity, Wireweed will undoubtedly have impacted on rock pool and shallow marine habitats but with little baseline data from before its arrival in the Channel Islands, the nature and scale of this impact can only be speculated. Bracken (2012) found that dense areas of Wireweed on Jersey's seashore had probably crowded out most other seaweed species but, conversely, might have provided shelter for small arthropods and fish. Few animals graze on Wireweed and evidence suggests that its impact on some high biodiversity habitats, such as seagrass beds, is minimal.

The economic effects of Wireweed are also not fully understood. Dense stands of seaweed can jam boat propellors and trap fishing tackle but its

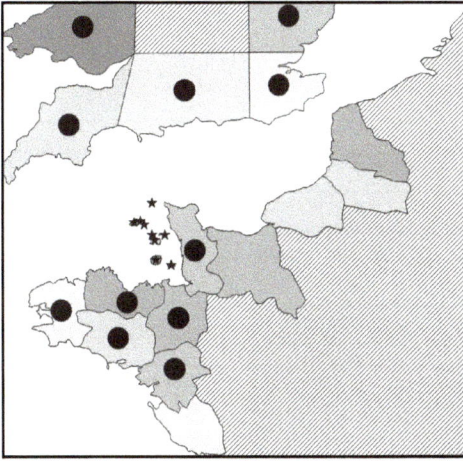

potential effects on aquaculture and commercial fishing within the islands has not been quantified. There is no evidence of any Channel Islands government having spent money clearing Wireweed from sites but in other parts of Europe the choking effect of the weed has been a menace to small harbours and shallow marine aquaculture areas.

Wireweed has probably reached its maximum extent within the Channel Islands and, as there is no known means of controlling or eradicating it, the species now forms a part of seashore ecology in the English Channel. It is sometimes used as a visible example as to why biosecurity is important when it comes to the movement of boats and aquaculture seed stock around the world.

*TL: Wireweed dominating a flooded gully area on Jersey's south-east coast. TR: A young plant in December 2016; Wireweed will usually die back in the late summer and autumn but new growth can begin in November. BL: Details of the leaves and seeds. BR: Floating plants (5 metres in height) photographed underater at Les Minquiers.*

# *Colpomenia peregrina*
## Oyster Thief

Threat score: 4

| ES | Hab | Tox | Econ |
|----|-----|-----|------|
| 2 | 2 | 1 | 1 |

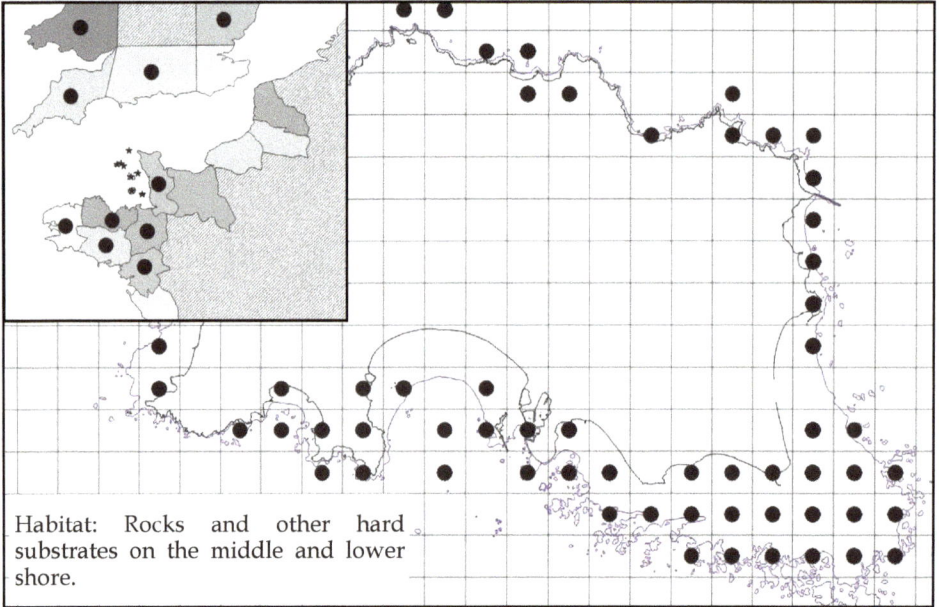

Habitat: Rocks and other hard substrates on the middle and lower shore.

A distinctive seaweed that is native to the Pacific Ocean but which was introduced into France in 1905 with oyster stock and has since spread widely to many parts of the European Atlantic coast. Commonly found on most Jersey coasts, *C. peregrina* has been established for a long time and it does not show signs of spreading rapidly. It is not thought to present a potential threat to local ecology of species.

*Above: Specimens of* Colpomenia peregrina *growing on Jersey's south-east coast.*

# *Dictyota cyanoloma*
## Blue Fringe Fan Weed

Threat score: 12

| ES | Hab | Tox | Econ |
|----|-----|-----|------|
| 3 | 4 | 1 | 1 |

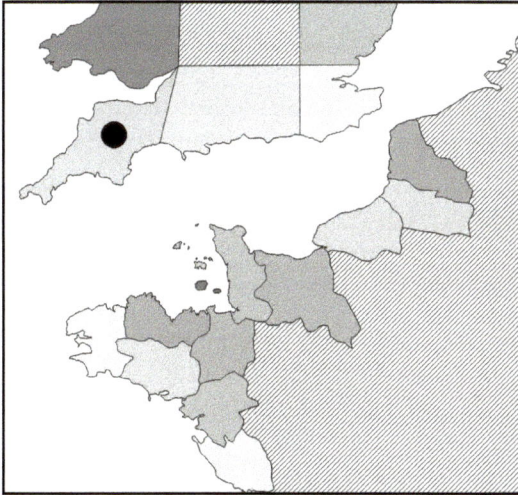

First described in 2010 from specimens found in the Mediterranean Sea, *Dictyota cyanoloma* was thought to have many characteristics associated with an introduced species. Recent work has found genetically similar populations of the seaweed in southern Australia suggests that it is non-native to Europe.

*D. cyanoloma* has spread rapidly and is now found on the Atlantic coast of Spain and, in 2013, on pontoons within Falmouth Harbour. It is largely subtidal but is thought to be able to spread into open coastal areas in the English Channel. *D. cyanoloma* is a species that has the potential to reach the Channel Islands and, being large and distinctive, should be included on a watchlist of potential non-native species when undertaking assessment surveys.

# Pikea californica
## Red Seaweed

Threat score: 1

| ES | Hab | Tox | Econ |
|----|-----|-----|------|
| 1 | 1 | 1 | 1 |

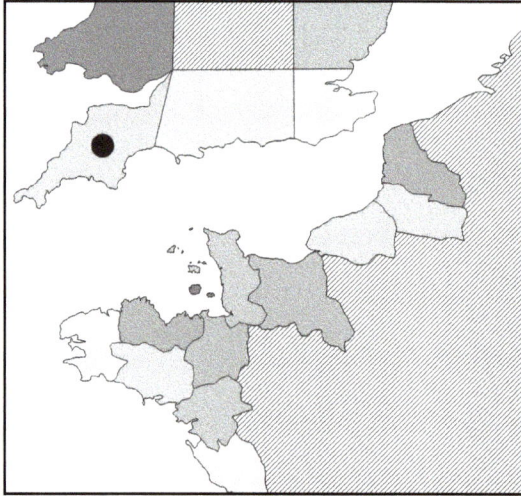

Known to have been present in the Scilly Isles since at least 1967, *P. californica* was recently disocvered on the Cornish mainland but has not so far spread any further in Europe. It does not seem to be spreading rapidly and is is unlikely to reach the Channel Islands but if it did then there is the potential for it to become established locally.

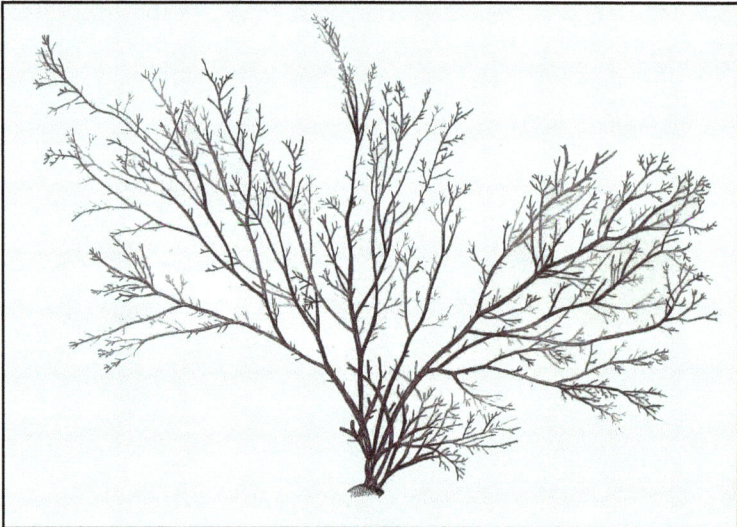

# Chrysymenia wrightii
## Golden Membrane Weed

Threat score: 4

| ES | Hab | Tox | Econ |
|----|-----|-----|------|
| 2  | 2   | 1   | 1    |

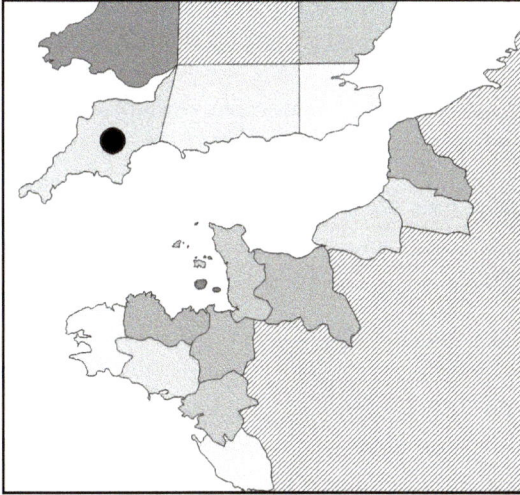

Native to Japan, *Chrysymenia wrightii* was first discovered in Europe in 1987 in the Mediterranean and then, in 2005, on the Atlantic coast of Spain. In 2013 specimens of *C. wrightii* were found growing on pontoons in Falmouth Harbour, Cornwall. This remains its only known English Channel location but it is probable that it will spread to other lcoations.

*C. wrightii* is a subtidal species that prefers sheltered conditions. As such, it has the potential to colonise harbours and marinas within the Channel Islands but may not be able to cope with the higher engery conditions on the open coasts and seas.

# *Sarcodiotheca gaudichaudii*
## Red Seaweed

Threat score: 4

| ES | Hab | Tox | Econ |
|----|-----|-----|------|
| 2  | 2   | 1   | 1    |

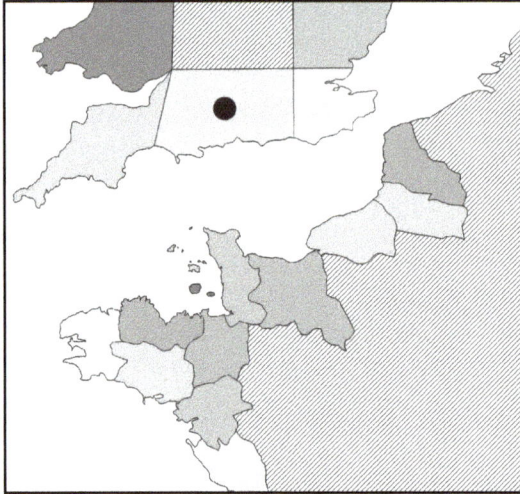

A red seaweed of the eastern pacific Ocean which was reported from the south coast of England in 1974. Further information has been difficult to obtain as the species is not listed in many standard reports on British non-natives. It would appear to have been reported from more than one locality but does not seem to be spreading rapidly. Prospects of it reaching the Channel Islands in the near future seem remote.

# *Bonnemaisonia hamifera*
## Red Seaweed

Threat score: 4

| ES | Hab | Tox | Econ |
|----|-----|-----|------|
| 2 | 2 | 1 | 1 |

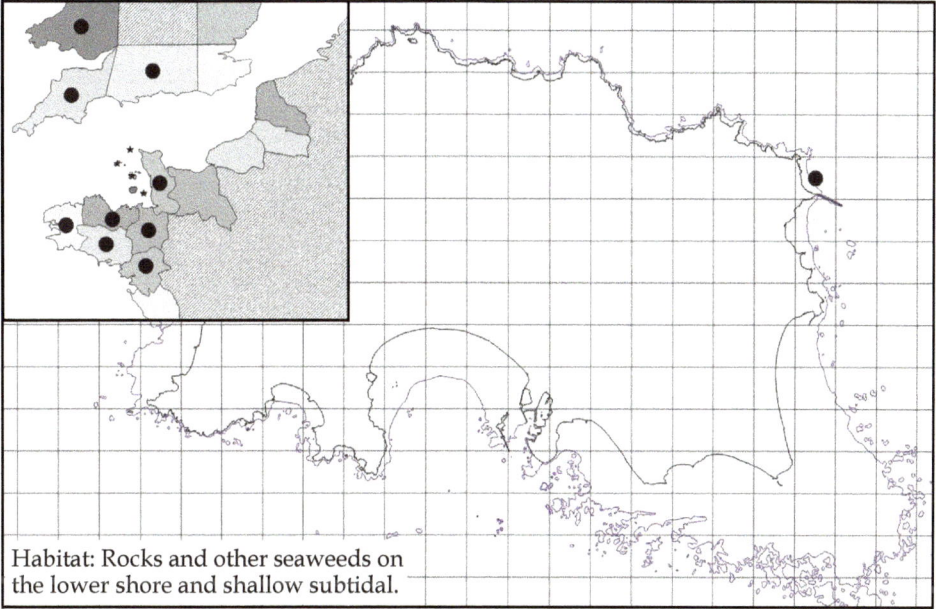

Habitat: Rocks and other seaweeds on the lower shore and shallow subtidal.

Native to Japan, *B. hamifera* was established on the Isle of Wight by 1894 and on the French coast by 1901. It was reported in Sark shortly afterwards and has probably been established locally since this time although the first Jersey record was not until 2003. Regarded as a fouling threat in some places, *B. hamifera* is rare in Jersey and not considered to be a potential menace.

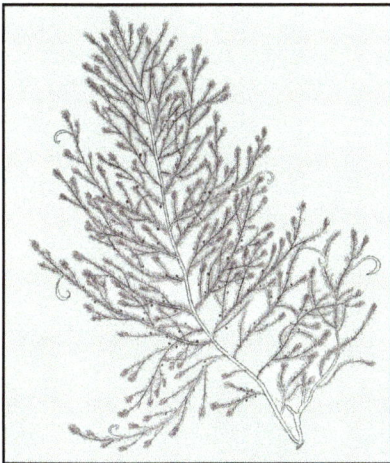

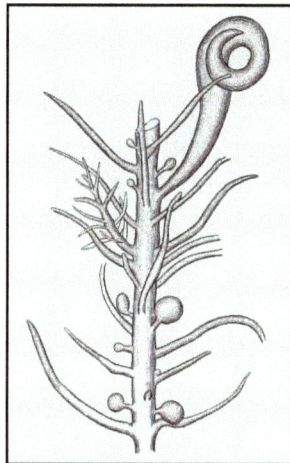

# *Asparagopsis armata*
## Harpoon Weed

Threat score: 24

| ES | Hab | Tox | Econ |
|----|-----|-----|------|
| 3 | 4 | 1 | 2 |

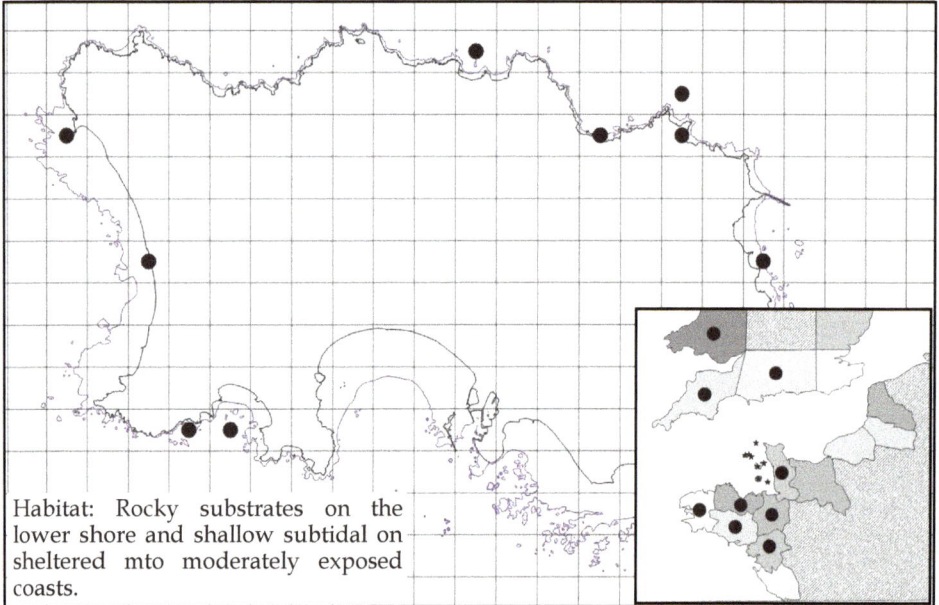

Habitat: Rocky substrates on the lower shore and shallow subtidal on sheltered mto moderately exposed coasts.

A distinctive red seaweed from the South-west Pacific which has been in European waters since the 1920s. *A. armata* has sexual and asexual phases which are morphologically very different and which have different distributions. Both phases are known from Jersey but it is generally the sexual phase that is reported as this is larger, colourful and more distinctive than the asexual phase.

A. armata has been known from the Channel Islands since the early 1950s and is common around Jersey but especially along the south-west coast. In most instances *A. armata* occurs in small aggregations but it has been observed to form a dense band on the lower shore at isolated locations. While not generally considered to be a serious threat to local habitats and species, *A. armata* can displace native species and have an effect on habitats, albeit localised. Casual monitoring of the species is recommended with any dense aggregations being of particular interest.

Right: *The sexual phase of* A. armata *in Bouley Bay, Jersey.* **Below:** *A microscope photo of one of the 'harpoons' on the sexual phase plant (x20).*

# *Grateloupia subpectinata*
## Fringe Weed

Threat score: 24

| ES | Hab | Tox | Econ |
|----|-----|-----|------|
| 3 | 4 | 1 | 2 |

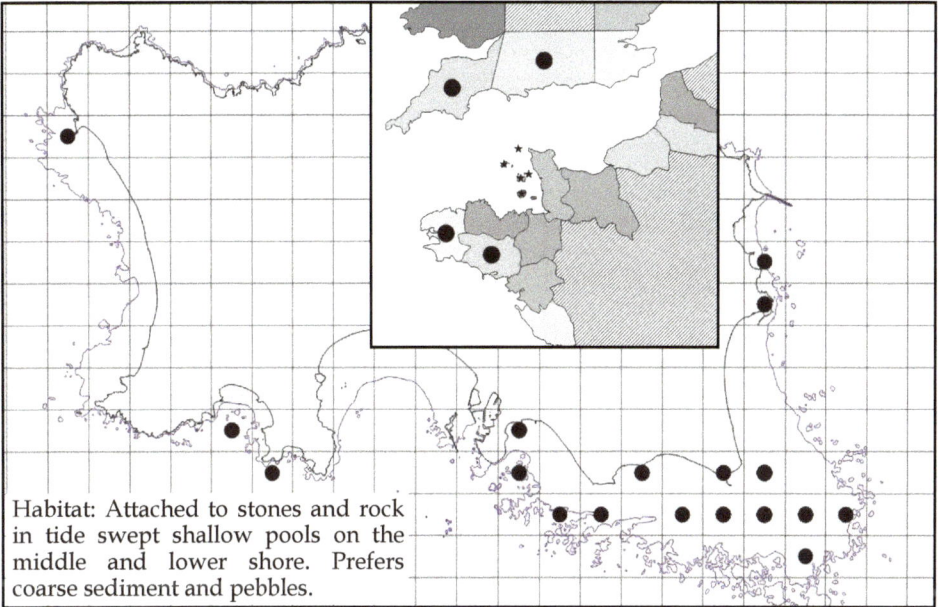

Habitat: Attached to stones and rock in tide swept shallow pools on the middle and lower shore. Prefers coarse sediment and pebbles.

Introduced into southern England from the Pacific Ocean in the 1940s, *G. subpectinata* has spread along many parts of the English Channel. It was probably established in the Channel Islands several years before the first report in 2011.

The distribution of *G. subpectinata* within Jersey is mainly along the south and east coasts where it may be abundant in shallow, wide rockpools on the middle and lower shore, especially those that cannot be colonised by *Sargassum muticum*. It appears to have become more widespread and abundant on the coast since in 2011 and may still be in an expansive phase. *G. subpectinata* has the potential to alter habitats and displace native seaweeds in the same manner as *Sargassum* did in the 1980s although, as a much smaller species, the effect will not be as severe. As such, *G. subpectinata* should be considered as a medium to high threat to local biodiversity and casually monitored.

*Left: Specimens of Fringe Weed growing in shallow pools near to La Rocque harbour, Jersey.*

175

# *Grateloupia turuturu*
## Devil's Tongue

Threat score: 24

| ES | Hab | Tox | Econ |
|----|-----|-----|------|
| 3 | 4 | 1 | 2 |

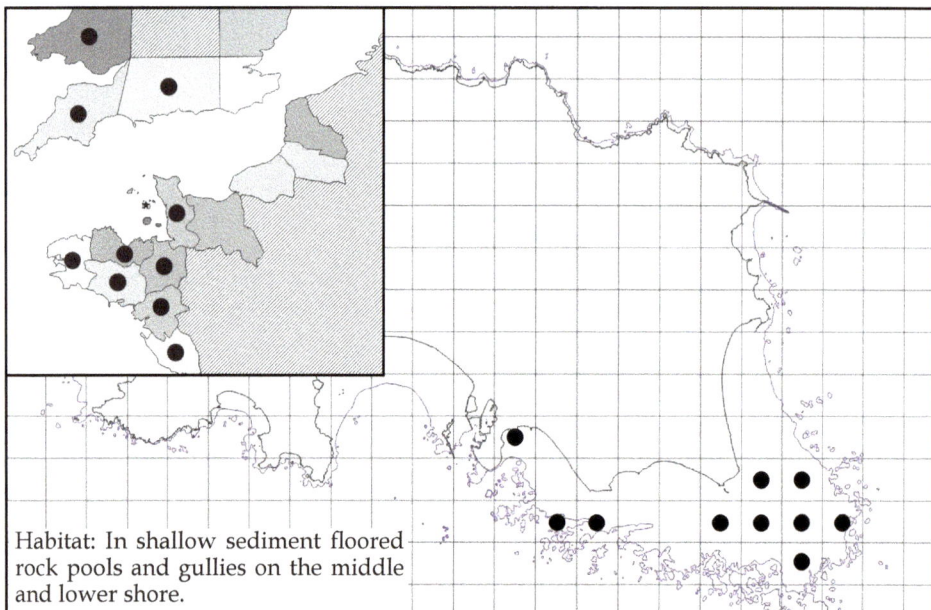

Habitat: In shallow sediment floored rock pools and gullies on the middle and lower shore.

Originally imported into Europe in the 1960s via aquaculture, G. turuturu was first discovered in the Normano-Breton Gulf in 1989 and then in Jersey *circa* 2000. It has not been reported from other Channel Islands but is probably present.

The plant is large and distinctive but has a localised distribution on Jersey, being generally found on the south-east coast. The population was low but stable until 2015 when it started to expand rapidly, in 2016 it had become one of the commonest seaweeds on the south-east coast and was starting to dominate some shallow pools. The effects on the local ecology are not known but while it is spreading, close monitoring is recommended.

Above: *Specimens of* G. turuturu *growing in rock pools on the south-east coast of Jersey.*

176

# Polyopes lancifolius
## Red Seaweed

Threat score: 4

| ES | Hab | Tox | Econ |
| --- | --- | --- | --- |
| 2 | 2 | 1 | 1 |

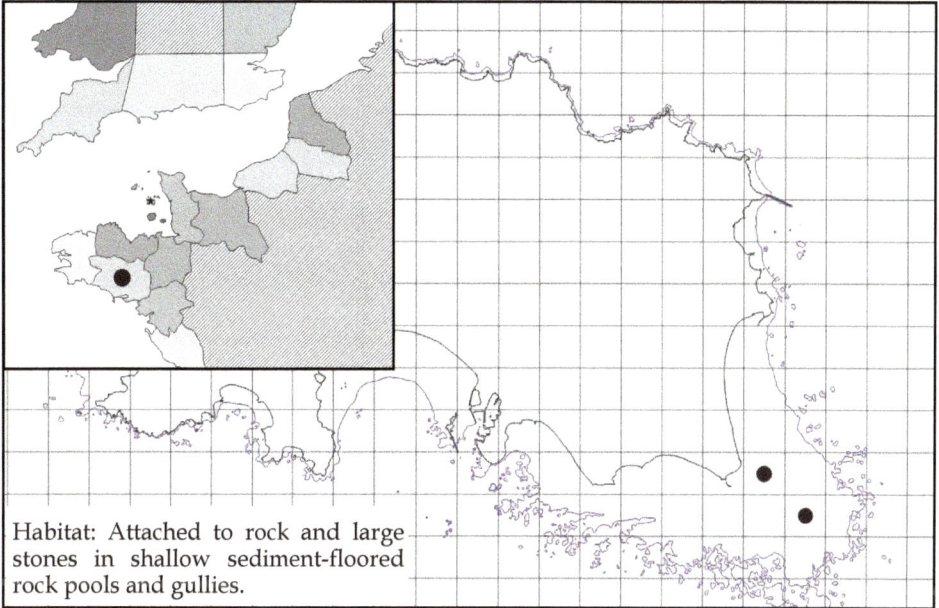

Habitat: Attached to rock and large stones in shallow sediment-floored rock pools and gullies.

The identification of *Polyopes lancifolius* on Jersey in 2011 is something of a mystery. This species is native to the North-west Pacific and was introduced into the north of Biscay in 2008 with aquaculture seed stock. However, it has remained localised to this area and there are no intermediary reports between Jersey and southern Brittany.

This suggests that *P. lancifolius* may have been introduced into the island via imported seed stock from southern Brittany which is against current biosecurity regulations on the island. As there is no evidence that unregulated seed stock has been imported, the occurrence of the seaweed remains unexplained. Three plants were found in 2011 followed by a further one in 2013. All were found in rock pools in the southern part of Grouville Bay. A lack of recent finds suggests it may be locally extinct.

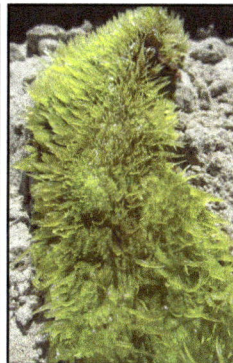

*Left: Specimens of P. lancifolius from Jersey. These are the 'hairy' phase of the plant which is prevelent in springtime. In the autumn and winter the leaves are smooth and shiny.*

# Solieria chordalis
## Red String Weed

Threat score: 12

| ES | Hab | Tox | Econ |
|----|-----|-----|------|
| 2 | 3 | 1 | 2 |

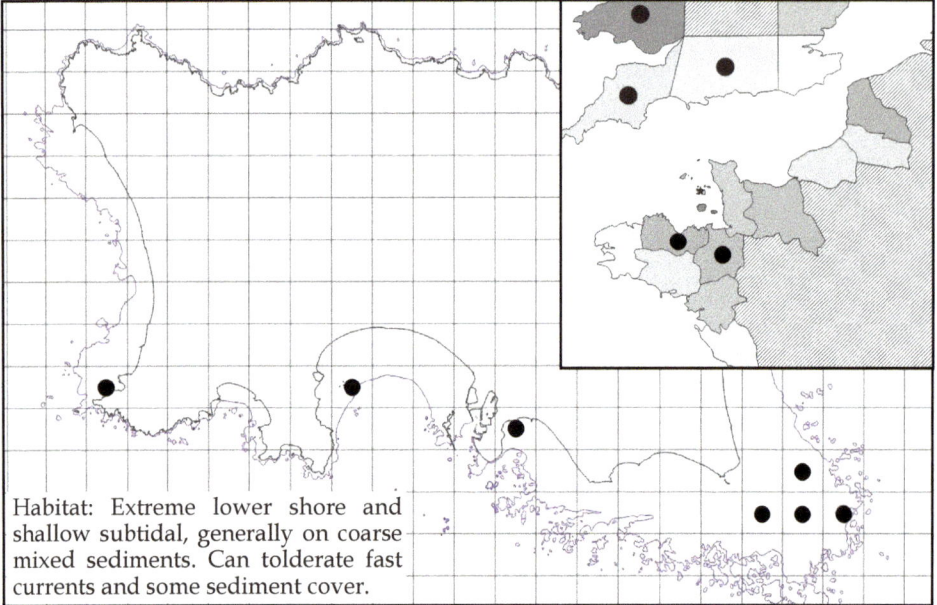

Habitat: Extreme lower shore and shallow subtidal, generally on coarse mixed sediments. Can tolderate fast currents and some sediment cover.

Native to the Mediterranean, *S. chordalis* is a large, distinctive red seaweed that was recorded in northern Brittany in 1964 and south-west England in 1976. It has since spread to several other locations but does not appear to be spreading rapidly within the region.

Although it can form localised dense stands, *S. chordalis* often occurs as isolated plants. The first Channel Island specimens were found in Jersey in October 2014 since when reports have been received of loose specimens and a few *in situ* plants. Recent evidence suggests that it is becoming more abundant and as a new arrival to the Channel Islands, this is a species that needs to be kept under close observation.

*Above: Specimens of* S. chordalis *from Jersey's seashore.*

# *Caulacanthus ustulatus*
## Red Seaweed

Threat score: 18

| ES | Hab | Tox | Econ |
|----|-----|-----|------|
| 3  | 3   | 1   | 2    |

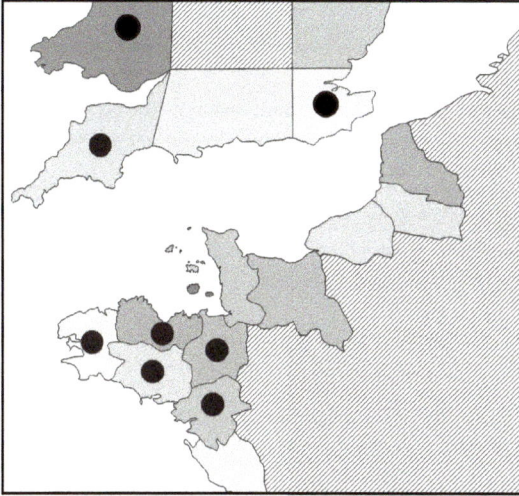

Introduced into southern Biscay from the North-west Pacific in the 1980s, *C. ustulatus* has spread to many other parts of the Atlantic coast and was recorded in the Normano-Breton Gulf in 2005. This is a short, turf forming intertidal species that has caused problems in some parts of the world by colonising large areas of the middle and upper shore rock, displacing native organisms such as barnacles, limpets and mussels. It can also potentially colonise aquaculture sites, presenting a potential economic issue.

The presence of this species on adjacent coasts and its habitat preferences suggests that it is (or will soon be) present in the Channel Islands. Recent searches on Jersey have proved negative but if *C. ustulatus* is foudn locally then it should be monitored for any potential effect on native habitats and species.

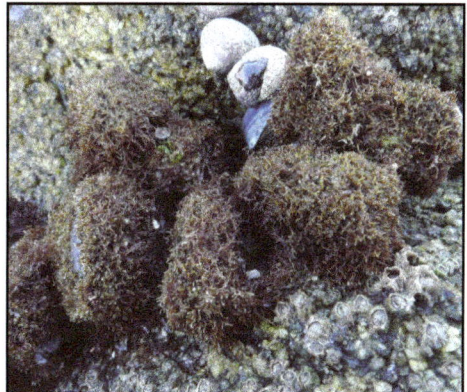

# *Gracilaria vermiculophylla*
## Red Seaweed

| ES | Hab | Tox | Econ |
|----|-----|-----|------|
| 2 | 3 | 1 | 1 |

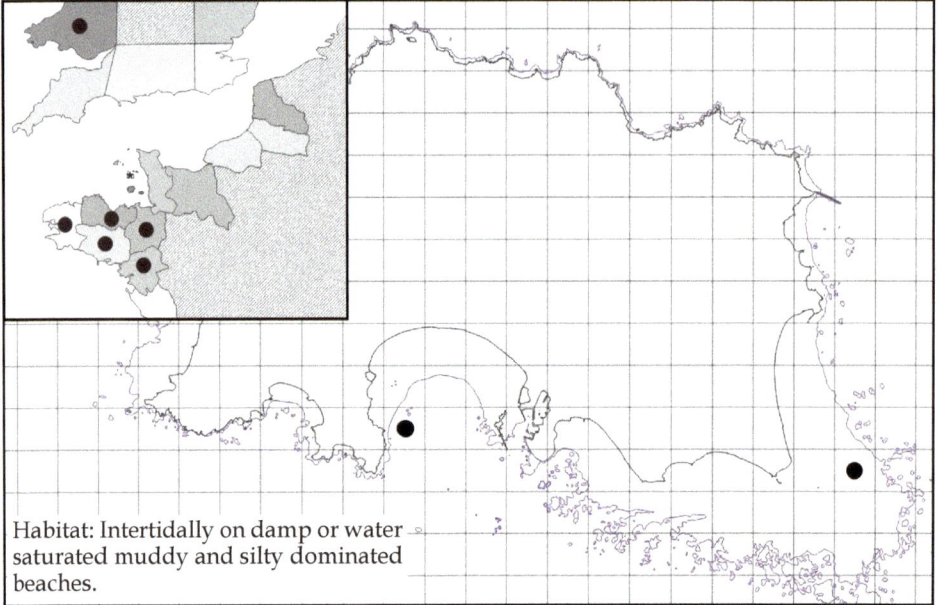

Habitat: Intertidally on damp or water saturated muddy and silty dominated beaches.

*Gracilaria vermiculophylla* is native to the North-west Pacific but was imported into the Bay of Biscay in the 1990s with aquaculture stock. It has spread along the Brittany coast and was first identified in Jersey in 2014. *G. vermiculophylla* has dominated some habitats in Brittany and is classed as a threat to local biodiveristy. In Jersey it is known from two locations from the south coast but an apparent preference for muddy habitats may limit its ability to spread widely. As a newly arrived non-native species, its progress should be monitored in the short to medium term.

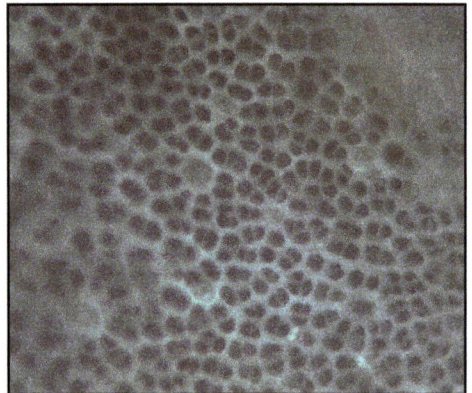

*Left: A specimen on muddy sand near to St Aubin's Fort, Jersey. Right: A section through the stem of the same specimen.*

# *Lomentaria hakodatensis*
## Red Seaweed

Threat score: 4

| ES | Hab | Tox | Econ |
|----|-----|-----|------|
| 2 | 2 | 1 | 1 |

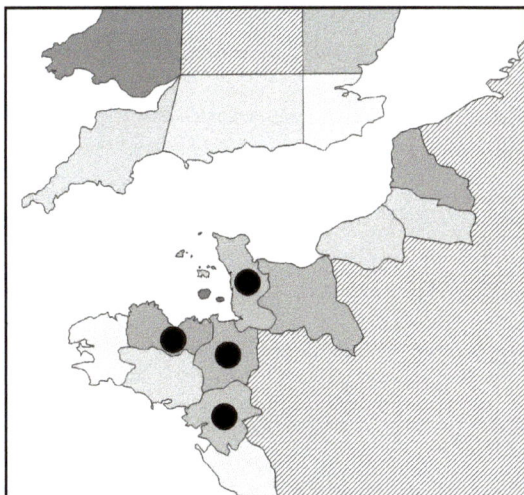

Native to the North-west Pacific, this red seaweed was recorded in southern Brittany in 1984 where it is thought to have been introduced via aquaculture. It has subsequently been reported from many other places along the Atlantic coast as well as the Mediterranean.

Although not reported from the Channel Islands (or the UK), *L. hakodatensis* has been recorded from North Brittany and the adjacent Normandy coast which makes it probable that it is either already established in the islands or will be soon. It is a relatively obscure seaweed which lives on the extreme lower shore and shallow subtidal and is often associated with the invasive seaweeds *Sargassum muticum* and *Undaria pinnatifida*. At present *Lamentaria hakodatensis* is not regarded as a serious threat to local habitats or species.

# *Aglaothamnion halliae*
## Red Seaweed

Threat score: 6

| ES | Hab | Tox | Econ |
|----|-----|-----|------|
| 2  | 3   | 1   | 1    |

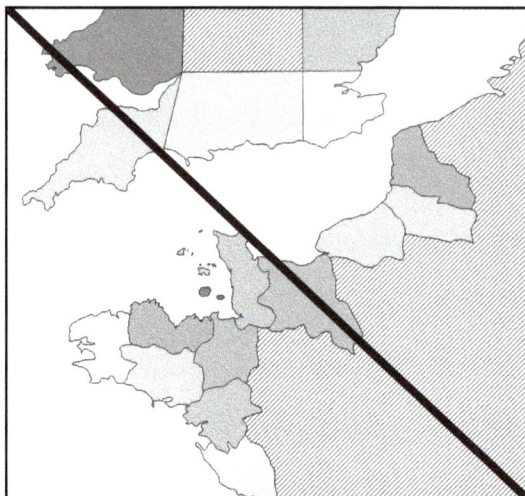

Native to the North-west Pacific, *A. halliae* was introduced into the Mediterranean in the 1990s and was then found in Norway in 2004. It has the potential to spread into the English Channel and is regarded as a medium threat to the British Isles where it may compete with native species.

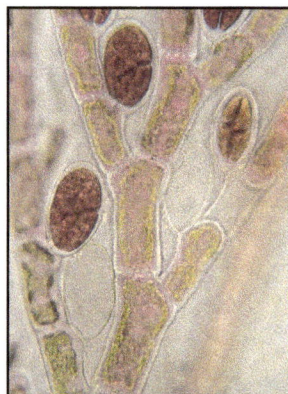

# *Anotrichium furcellatum*
## Red Seaweed

Threat score: 4

| ES | Hab | Tox | Econ |
|----|-----|-----|------|
| 2  | 2   | 1   | 1    |

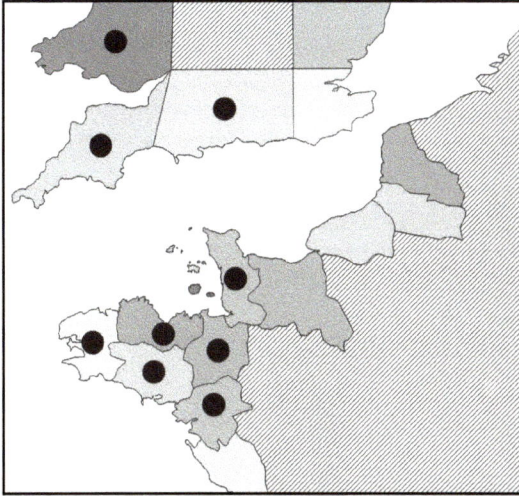

A delicate, small seaweed associated with fully marine intertidal bays. It is native to the North Pacific and was first found in Europe in 1922 at Cherbourg but by the 1950s was widespread along the Normandy and Brittany coasts. The widespread regional nature of this species suggests that it is probably present in the Channel Islands but, due to its size and confusion with other species, it is difficult to identify and may have been overlooked.

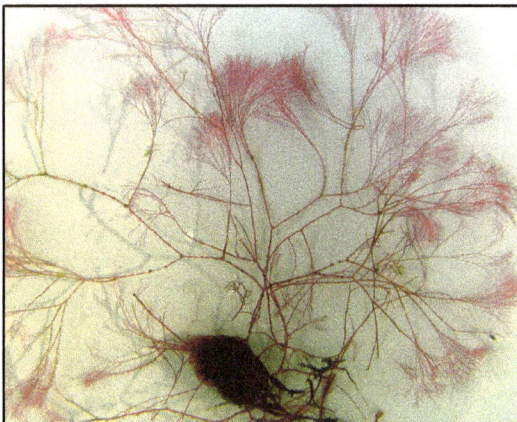

# *Antithamnion densum*
## Red Seaweed

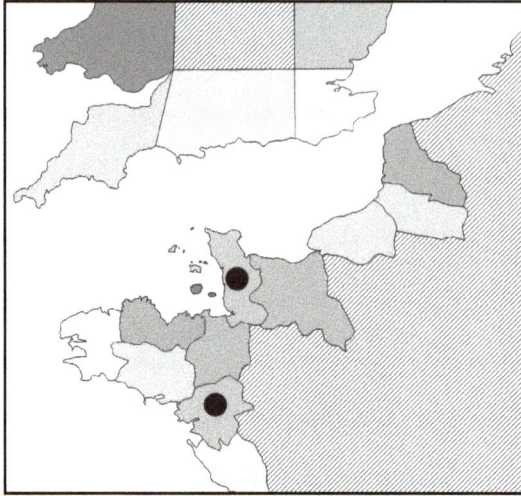

A seaweed from the Pacific coast of South America that was recorded in France in 1964 but by the 1990s had been recorded along much of the Atlantic coast of Europe. It is not listed as a threat and may already be present in the Channel Islands.

# *Antithamnion nipponicum*
## Red Seaweed

Threat score: 4

| ES | Hab | Tox | Econ |
|----|-----|-----|------|
| 2  | 2   | 1   | 1    |

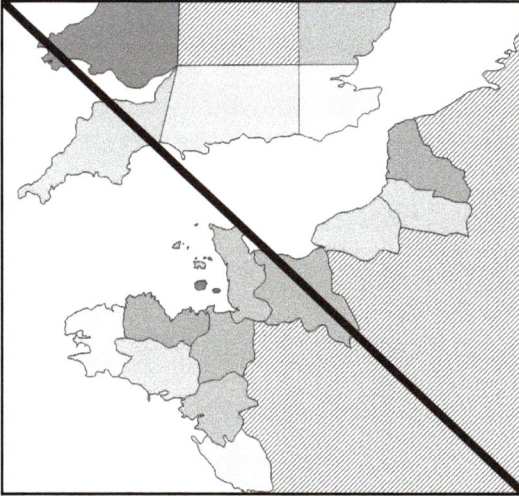

# *Antithamnion pectinatum*
## Red Seaweed

Threat score: 9

| ES | Hab | Tox | Econ |
|----|-----|-----|------|
| 3  | 3   | 1   | 1    |

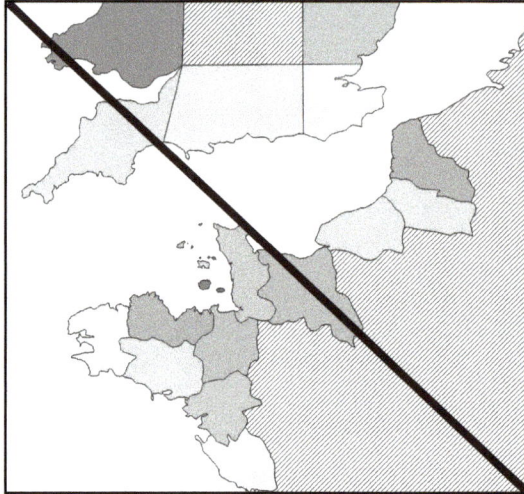

A North-west Pacific seaweed that was reported from the French Mediterranean coastline in 1988 but which during the 1990s spread to several other locations. The survey for this report did not find any records from northern Europe but the species is listed as a medium threat to the British Isles by Roy *et al.* (2014).

# *Antithamnionella spirographidis*
## Red Seaweed

Threat score: 8

| ES | Hab | Tox | Econ |
|----|-----|-----|------|
| 2  | 2   | 1   | 2    |

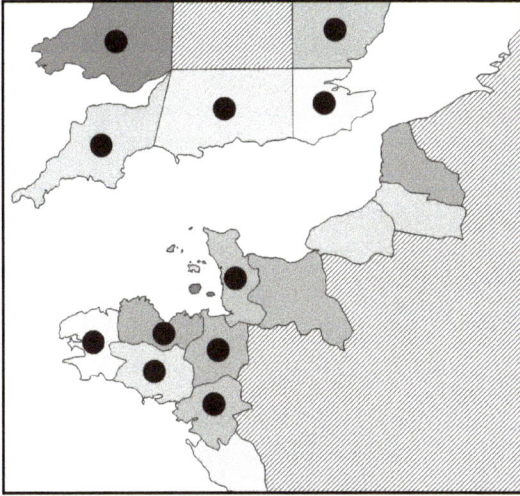

Native to the North-west Pacific and first reported from Cherbourg in 1927, *A. spirographidis* has since spread to many parts of the English Channel and Atlantic coasts. Although listed as a potential fouling species, *A. spirographidis* is not generally regarded as a threat. Based on its existing range and habitat preference, it is likely to be established in the Channel Islands already.

# Antithamnionella ternifolia
## Red Seaweed

Threat score: 4

| ES | Hab | Tox | Econ |
|----|-----|-----|------|
| 2 | 2 | 1 | 1 |

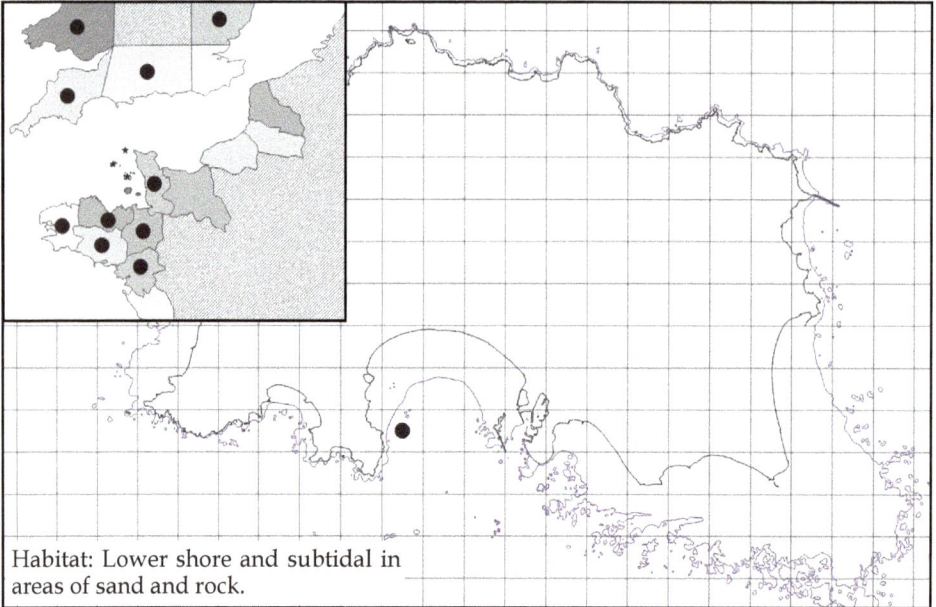

Habitat: Lower shore and subtidal in areas of sand and rock.

Native to the South Pacific, the first European record for *A. ternifolia* was in 1906 from Plymouth but it has since become widely distributed around the British Isles and Atlantic coast to Portugal. The first Channel Island records are from the 1930s in Guernsey where this alga was mistaken for a new species (named *A. sarniensis*) and a specimen was found in Jersey in 2011. Although noted as a fouling species in some areas, it is not common in the Channel Islands and is not regarded as a potential threat.

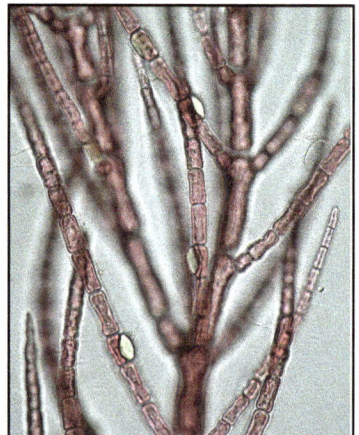

# *Spongoclonium caribaeum*
## Red Seaweed

Threat score: 4

| ES | Hab | Tox | Econ |
|----|-----|-----|------|
| 2  | 2   | 1   | 1    |

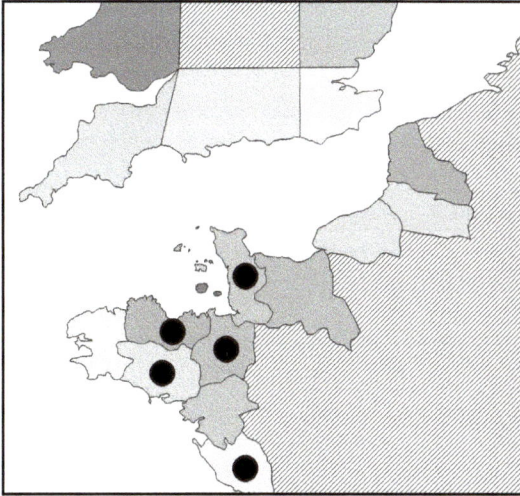

A cryptogenic species that was first described from the Caribbean but which has been reported from many other regions worldwide including the Atlantic coast of Europe where it has been reported from at least the 1960s onwards. It is not listed as a serious threat and has been widely reported from the Normano-Breton Gulf. Although not yet reported from the Channel Islands, it is probable that the species is (or will shortly be) reported from this area.

# *Dasysiphonia japonica*
## Red Seaweed

| ES | Hab | Tox | Econ |
|----|-----|-----|------|
| 4  | 4   | 1   | 1    |

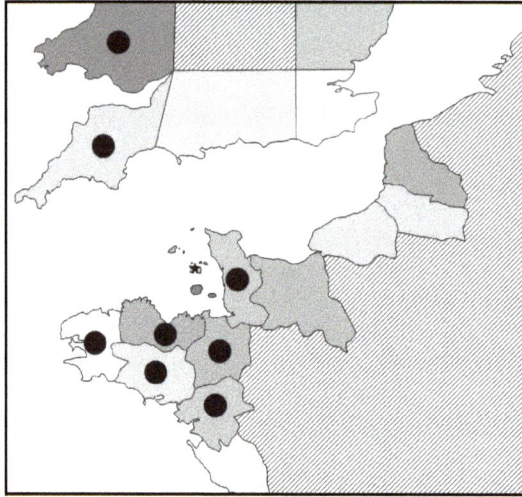

First recorded in 1984 from Roscoff, this red seaweed is native to the North-west Pacific and has since spread widely through the English Channel reaching the UK in 1999. Since 2005 *H. japonica* has increased markedly in abundance along some parts of the north Brittany coast but especially near Corbeau where plant densities went from 4 to 174/m² in just a few years.

It is a subtidal species that can cope with a range of conditions. It seems to be rarely reported in the Channel Islands but this could because of its subtidal nature and similarity to other species. Given issues with excessive growth in other parts of the world, including Brittany, it is a species that needs to be monitored.

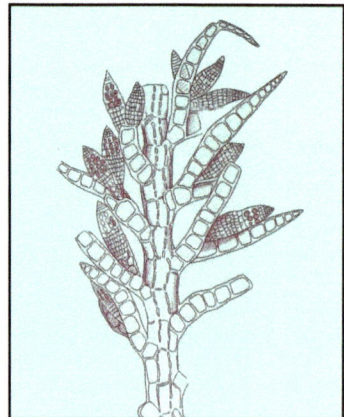

# Laurencia brongniartii
## Red Seaweed

Threat score: 4

| ES | Hab | Tox | Econ |
|----|-----|-----|------|
| 2 | 2 | 1 | 1 |

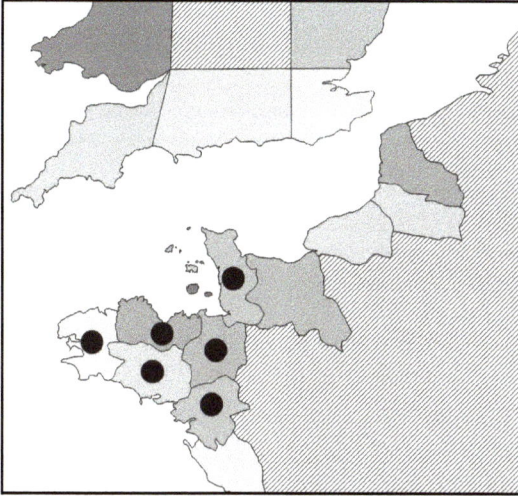

Native to the North-west Pacific but reported from Brest in 1989, *L. brongniartii* has since had specimens identified in Normandy, Ireland and Spain although not in the UK yet. It is not regarded as a serious threat to the local ecology and may already be established in the Channel Islands. Several similar species occur locally and it may be difficult to identify *L. brongniartii* from these.

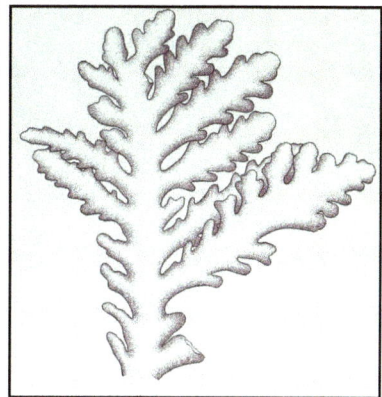

# *Neosiphonia harveyi*
## Red Seaweed

Threat score: 8

| ES | Hab | Tox | Econ |
|----|-----|-----|------|
| 2 | 2 | 1 | 2 |

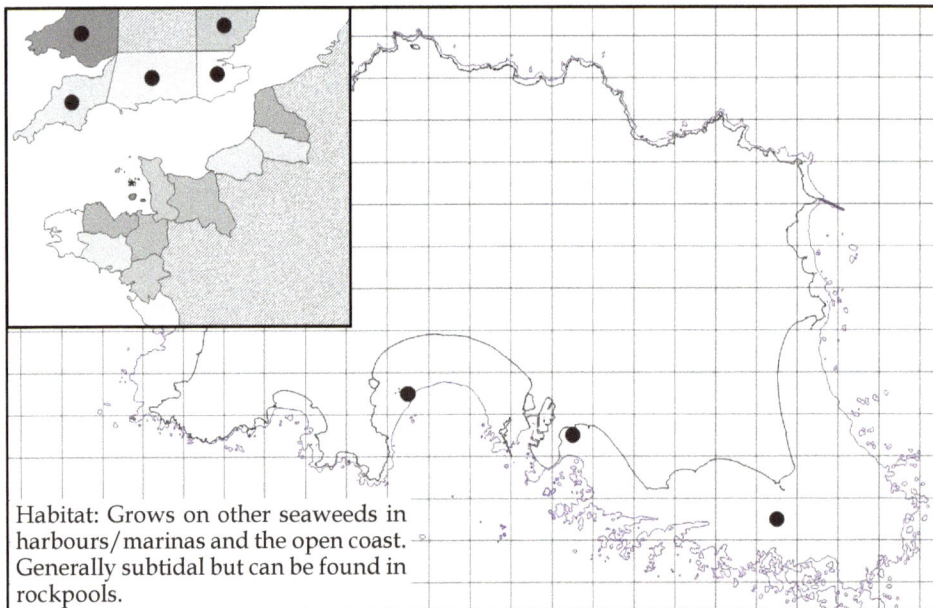

Habitat: Grows on other seaweeds in harbours/marinas and the open coast. Generally subtidal but can be found in rockpools.

A large feathery species of red seaweed that may have been present in Europe as early as 1832 and certainly by 1908 when it was found in the UK. It is widespread and can occur in abundance especially in harbours and ports where it may become a fouling menace. In Jersey it has been recorded from the seashore but is not known from the harbour or marina areas. In February 2017 a rapid assessment survey in the marina at Granville (Normandy) found it to be common on the pontoons.

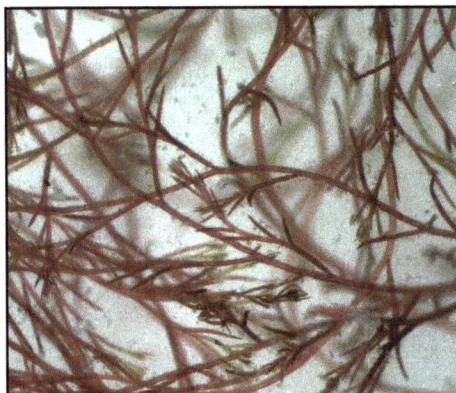

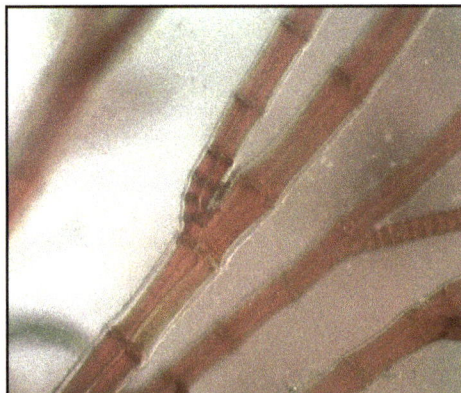

Left: *A specimen of* N. harveyi *taken from Granville marina (Normandy) in February 2017.* Right: *The same specimen under a microscope.*

# *Cryptonemia hibernica*
## Irish Thread Weed

Threat score: 4

| ES | Hab | Tox | Econ |
|----|-----|-----|------|
| 2 | 2 | 1 | 1 |

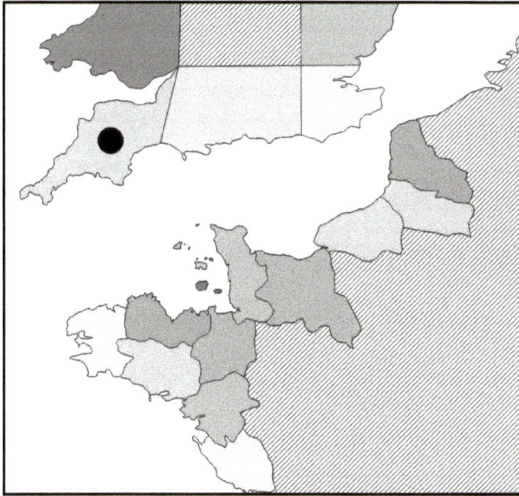

Specimens of *Cryptonemia hibernica* were first discovered and described in Cork Harbour, Ireland, in 1971. Its restricted distribution and biological similarity to other *Cryptonemia* species found in the eastern Pacific Ocean give it the characteristics of an introduced species and it is presumed to be non-native.

At present its known European distribution is from isolated locations in southern and northern Ireland and from Plymouth sound. It is generally (but not exclusively) subtidal with a preference for sheletered areas and kelp forests. The potential for *C. hibernica* to reach and establish itself in the Channel Islands is not known.

Left: *A specimen growing on a rocky shore in Ireland.* Right: *a herbarium specimen.*

# *Caulerpa taxifolia*
## Killer Algae

Threat score: 12

| ES | Hab | Tox | Econ |
|----|-----|-----|------|
| 3 | 4 | 1 | 1 |

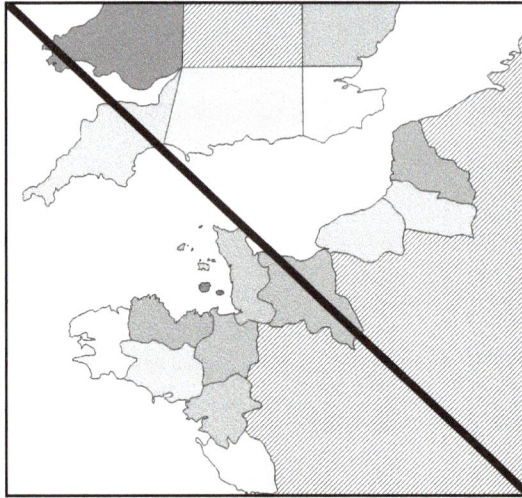

An Indian Ocean species that was accidentally introduced into the Mediterranean in 1984 after being released from an aquarium. Although not yet in northern Europe, it is believed to have the potential to tolerate conditions in the English Channel and is regarded as a medium risk to the British Isles. There are concerns that *C. taxifolia* will be transported to North-west Europe where it could dominate shallow marine environments as it has in the Mediterranean, displace native species.

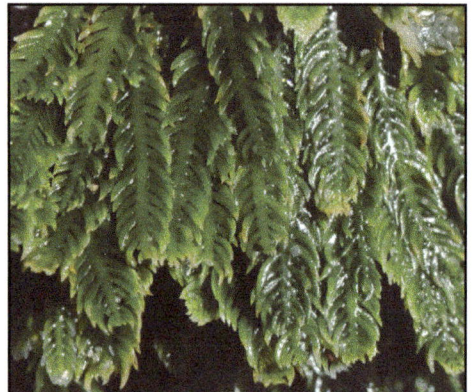

Left: *A meadow of* C. **taxifolia** *off the western USA coast.* **Right**: *Detail of an individual plant growing on Lord Howe Island, Australia.*

# *Codium fragile* var. *fragile*
## Velvet Horn

Threat score: 12

| ES | Hab | Tox | Econ |
|----|-----|-----|------|
| 2 | 3 | 1 | 2 |

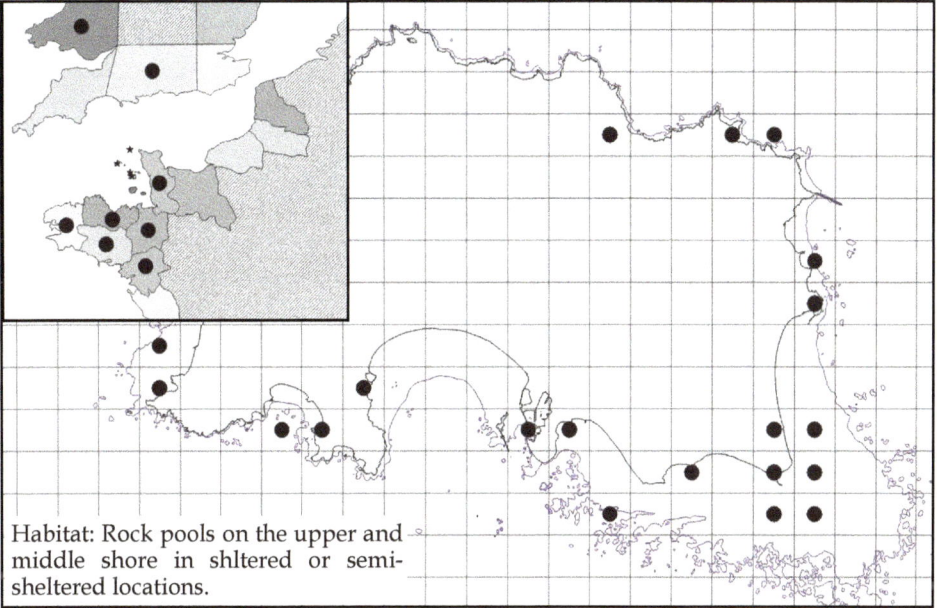

Habitat: Rock pools on the upper and middle shore in shltered or semi-sheltered locations.

There has been some debate as to which species and subspecies of Codium should be classed as non-native. Here we have taken the opinion of Brodie et al. (2007) who class *C. fragile* var. *fragile* as non-native but not var. *atlanticum* (which is considered non-native by other authors).

*C. fragile* var. *fragile* is common on Jersey and has probably been here, and in other Channel Islands, for many years. Identification is an issue as it can only be secured using a microscope or powerful hand lens. This means it is probably under-recorded across all the islands.

Left: *A specimen of* C. fragile *var.* fragile *at La Rocque, Jersey.* Right: *The same specimen viewed under the micrscope showing its pointed utricles.*

195

# - Part Four -

## Summary and Recommendations

# 6 - Summary and Conclusions

The results from this project suggest that when it comes to non-native species, the Channel Islands are in an unusual situation compared with most European coastal areas.

The geographical location of the Channel Islands places them in a bioprovince which has historically received a large number of non-native species, especially from the North-east Atlantic and North-west Pacific. The Channel Islands are also located midway between two hubs where a disproportionate number of non-native species seem to have entered Europe. One hub is the coast of southern England, where species have arrived in large harbours such as Southampton through international shipping. The other hub is the aquaculture area of northern Biscay, where non-native species have entered Europe (and then translocated) with seed stock.

The Channel Islands are connected to these areas via a shallow seabed and local tidal currents but they also receive commercial and leisure shipping from their ports. This provides both natural and artificial means by which non-native species disperse away from their point of origin to establish themselves at other locations, including the Channel Islands.

The net result of this is that the Channel Islands can potentially receive non-native marine species that initially entered Europe at locations some distance to the north and south. This places the islands at a crossroads between the English Channel and the Bay of Biscay making them potentially accessible by a high number of non-native species in comparison with many other European coastal sites. However, the number of non-native species which could colonise the Channel Islands is tempered by physical parameters (such a sea temperature and tidal currents) and the absence of certain marine habitats. Particularly relevant is a lack of any brackish water areas, tidal mud flats and saltmarshes which means that the many non-native species which have colonised European estuaries (such as the Chinese Mitten Crab) will be absent from the islands.

A desktop survey of information sources on non-native marine species within the geographic area covered by the Channel Islands produced a shortlist of 134 species that have the potential to reach the islands and establish themselves. A further survey of local records and fieldwork found that there have been 43 established non-native species reported in the islands but that there may be a further 25 species which are established but have yet to be identified.

Each of the shortlisted 134 species has been assessed for its biology, behaviour and potential impact on the environment, health and economy of the Channel Islands. A summary of the main findings from this survey and assessment is given in this chapter but reading of the whole report is recommended in order to understand fully the issues and information presented here.

## 6.1 - Hotspots and Hubs

During survey work on Jersey it was noted that several locations appeared to have had a disproportionate number of non-native species reported in comparison to neighbouring coastal areas. Such hotspots (also known as hubs) are a familiar feature with non-native species and they often represent locations where species can easily be transported (e.g. harbours) or where the local environment is conducive to the establishment of certain species (e.g. power station outfalls).

Distribution data from the Channel Islands suggests that the marinas at St Helier and St Peter Port are home to a large number of non-native species. These 'hubs' contain up to ten non-native species most of which live on artificial structures such as pontoons, buoys, ropes, etc. Several appear to be restricted to the marinas (e.g. *Tricellaria inopinata*) or were initially only known from a marina before later spreading into the wider marine environment (e.g. *Watersipora subatra* and *Undaria pinnatifida*). It is assumed that non-native species exhibiting these characteristics arrived in the islands at their marinas, probably as a result of shipping movements from France and the UK.

That the marinas should be hubs for non-native species is not unusual. Surveys in the UK and elsewhere highlight the importance of harbours and marinas in the establishment and spread of non-native species (see especially the work of Tidbury *et al.*, 2014).

*Figure 6.1 - A specimen of Wakame (*Undaria pinnatifida*) growing amongst a mass of smaller encrusting animals and plants on a pontoon at La Collette marina, St Helier.*

Foster *et al.* (2016) studied multiple harbours in the UK and, according to their results, the marinas at St Helier exhibit many characteristics that favour the establishment and maintenance of non-native species. These include semi-enclosed entrances, floating pontoons, little freshwater influence and a mix of internal concrete seawalls and boulder breakwaters. The periodic monitoring of Channel Islands' marinas (probably via rapid risk assessment surveys) will be important if the arrival of a new non-native species is to be detected promptly.

Another possible hub area is at La Collette, Jersey, immediately to the east of the island's energy from waste plant. The seashore surrounding the plant's water outfall has had a high number of non-native species recorded from its vicinity. This includes some species, such as *Watersipora subatra*, that had previously only been known from St Helier's marinas. The water intake for the power plant is situated inside St Helier Harbour and it is possible that this be a means by which species can leave the harbour. However, the elevated water temperature associated with the La Collette outfall might also make it easier for some non-native species to establish themselves there (e.g. see discussion in Ryland *et al.*, 2011).

The area of seashore to the south-east and west of La Rocque Harbour has a large number of non-native seaweed species present some of which are either not found elsewhere or are proportionately much rarer. Species that are common in this area include *Grateloupia subpectinata*, *Undaria pinnatifida* and *Grateloupia turuturu*.

It is possible that a high concentration of biotopes in this area (especially rock pools and flooded gullies) provides a more suitable habitat than the exposed, rocky shores found along the north and west coasts. This area is adjacent to Jersey's intertidal aquaculture production areas but, with the exception of *Ruditapes philippinarum* and possibly *Polyopes lancifolius*, this is not thought to be the means by which the species arrived in the area. Most of the species concerned seem to have dispersed naturally from the adjacent French coasts.

*Figure 6.2 - Flooded gully systems on Jersey's south-east coast seem to be particularly attractive to non-native species and especially seaweeds.*

The identification of hub locations is useful as these areas can be monitored regularly so that newly arrived non-native species can be picked up quickly. At present the most detailed Channel islands distribution data for non-native species is from Jersey, which is why most of the known hubs are on that island. As more detailed data become available from other islands, so it should be possible to identify other hub areas. All hub areas should be subject to a regular programme of rapid assessment surveys and the results fed back to the main biological recording facilities in the islands, UK and France.

## 6.2 - Data Coordination and Dissemination

Studies on non-native species have tended to occur at a regional or national level with few of the conclusions being coordinated or feeding into geographically wider, coordinated projects or databases.

The Channel Islands are located at a crossroads between southern Europe and the English Channel (see Chapter 4.1) which means any assessments and horizon scanning for non-native species has to take place across a wide geographical region. This requires looking at survey and research work from both sides of the English Channel, the southern North Sea, the Normano-Breton Gulf and the Bay of Biscay (see Figure 2.3).

During the information gathering phase of this study, no single source was able to provide a coherent non-native species list covering the Channel Islands region. (A possible exception is Goulletquer, 2016, which was published while this report was being prepared.) Information was spread across a wide range of published and unpublished sources and researching the origin, occurrence, distribution, biology and behaviour

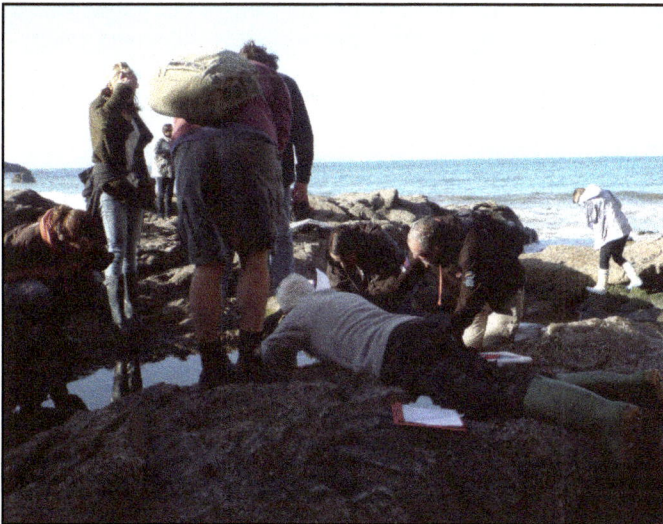

*Figure 6.3 - Members of the Société Jersiaise surveying the seashore at Grève de Lecq, Jersey. This work has produced many non-native species records.*

of individual non-native species often required extensive literature and database searches followed by cross-referencing and checking. Regional and national databases, such as GBNNSS and NBN Gateway often had good information about the local distribution and behaviour of particular species but a number of non-native species known to occur in the region have no records.

This experience agrees with the conclusions of Sambrook *et al.* (2014) and Stebbing *et al.* (2014) who found that, while valuable baseline research has been undertaken on many European non-native species, much of the resultant information and conclusions is not being fed upwards into regional, national or international databases and other information sources. Having up-to-date national (or European) registers and databases of non-native species will be a valuable asset when it comes to assessing and tackling this issue at a local level.

The Channel Islands has good biological recording centres in Guernsey, Jersey and Alderney but they do not have a centralised reporting structure for non-native species. There is also no coordinated research or monitoring within or between the islands. This should be addressed with all information being made public via the internet but also through the submission of records to other national projects and databases such as NBN Gateway and GBNNSS.

## 6.3 - Monitoring and Reporting

The identification of most non-native species in the Channel Islands has occurred within the past three decades, something which probably reflects renewed local interest in marine biology since the early 1980s. However, monitoring and reporting across the islands is at present irregular. The patchy identification and recording of marine non-native species is an issue that faces most coastal locations in Europe. This is principally due to a lack of local specialist knowledge and/or regular monitoring both of which are required to obtain a full understanding of the diversity and behaviour of local non-native species.

A comparison of the taxonomic composition of non-native species known from the Channel Islands against those from the wider region suggests that several groups are being under-recorded locally. This includes phytoplankton, annelids, barnacles, sea squirts and red seaweeds, most of which require specialist knowledge to facilitate identification. Groups that have a good track record of identification within the Channel Islands include molluscs, large crustaceans and brown seaweeds, most of which are relatively easy to spot and identify by fishermen and amateur naturalists.

Central and local government often does not have the resources to pay for regular coastal monitoring and so this task is frequently left to NGOs and amateur naturalists. While the work of NGOs and naturalists is often effective, a lack of coordination and understanding between government, groups and individuals sometimes means that records of non-native species are not given sufficient consideration or do not become incorporated into regional or national datasets. However, in recent years there have been attempts at creating nationwide monitoring and reporting projects such as the Marine Biology Association's Shore Thing and smartphone apps such as Sealife Tracker.

The Channel Islands has a tradition of nurturing good networks of amateur naturalist networks particularly through the Société Jersiaise and Société Guernesiaise and, more recently, through NGOs such as the Alderney Wildlife Trust and National Trust for Jersey. These networks often operate with minimal help from government and do not coordinate on non-native species reporting although the Société Jersiaise and Société Guernesiaise do publish selective records annually.

The inaccessibility of local records is an historical issue which was largely alleviated by the creation of the GBRC and JBC. These have provided a centralised reporting and querying structure for Channel Island biological records which includes non-native species. Neither of these centres has a separate recording or reporting structure for non-native species but developing this should be relatively straightforward. Additionally, it may be that the biological records centres and NGOs such as the SJ and SG can be used to gather information via organised surveys and projects.

The use of rapid assessment surveys (RAS) for non-native species at selected locations (such as harbours and marinas) has been favoured in recent years as this can utilise local NGOs and naturalists to deliver good baseline information. The implementation of regular RAS for a handful of locations around the Channel Islands, possibly organised through one or more NGOs, will provide baseline data on easy to identify non-native species. This could be supplemented by more specialist work perhaps via student projects or encouraging/funding field studies by off island experts or specialist societies. There have been some organised national projects relating to marine non-native species with which the islands could participate if time and resources permit it.

The key with establishing RAS and other studies is to ensure that any information gathered is returned promptly and incorporated into reporting structures that can be accessed by government, NGOs and the wider world. In this respect, the ultimate destination for any RAS data should be one of the islands' biological records centres. Regular species' assessments should be made on the offshore reefs as these can act as 'control areas' in comparison to the coastal regions of the bigger Channel Islands.

Additional measures could include devising a non-native species strategy for all port and harbour areas within the Channel Islands. This would aim to minimise the establishment and spread of non-native species through a code of conduct regarding the movement and cleaning of infrastructure and awareness raising among resident and visiting yachtsmen. Additionally, the routine inspection of pontoons and offshore navigation buoys for non-native species when they are removed from the water is essential (Tidbury et al., 20 15). This would require the harbour authorities alerting others to the imminent removal of such infrastructure.

*Figure 6.4 - The removal of a pontoon float from St Helier Marina in 2011. Cooperation from Ports of Jersey meant that specialists were able to examine the float on the quayside and make a list of the species (native and non-native) attached to it. Such cooperation is vital if future monitoring is to be effective.*

Currently, there is a summary of the status of Jersey's marine non-native species published yearly by the Marine Resources Section. This practice should continue and the summary could perhaps be made more widely available by its publication on the GBRC, JBC or Société Jersiaise website. Any opportunity to share fieldwork or data with other organisations inside and outside of the island should be seriously considered.

## 6.4 - Economic, Social and Environmental Impact

The effect that non-native species can have on terrestrial areas may be very obvious. For example, in north-west Europe introduced species such as the Grey Squirrel, Japanese Knotweed and the Asian Hornet have had major effects on local species, habitats and landscapes.

Some marine non-native species have the potential to be highly disruptive but the inaccessibility of the marine realm can mean that their effects are hard to notice. In the Channel Islands the most obvious non-native species is Wireweed (*Sargassum muticum*) whose ubiquity, ability to grow to several metres in length and to choke rock pools and channels makes it unmissable. Below the waves divers will be familiar with the American Slipper Limpet (*Crepidula fornicata*) which can cover 100% of the seabed in some areas. However, many other high impact species are less noticeable as they may live under stones or remain unrecognised by all but a handful of knowledgeable naturalists and experts.

Judging the likely effect of a newly arrived non-native species is not straightforward. As with so many areas of the environment, by the time the scale of a threat is realised it is often too late to do anything about it. For this reason the last decade has seen the development of schemes which try to quantify potential threats posed to local areas by both established non-native marine species and those that might arrive in the following years.

204

This report adapted one of these threat scoring schemes and used it to rank the likely effect of 134 non-native species on the Channel Islands. Each of these species was assessed using four main categories concerning their impact on ecosystem functions, habitat/species, health and economy. Most of the non-native species were deemed to pose a low to moderate threat but 33 of them are thought to be of more concern.

Previously concern about non-native species within the islands has been focused on organisms that have the potential to affect economic activities such as aquaculture (e.g. *Bonamia ostreae*). It is only recently that the threat posed by other organisms, such as the Slipper Limpet (*Crepidula fornicata*), has been recognised locally. Although it is hard to quantify the monetary cost of non-native species, some do have the ability to impact heavily on the fishing and aquaculture industries. This includes the Slipper Limpet, the Veined Rapa Whelk (*Rapana venosa*) and several types of toxic pathogen and phytoplankton.

Other economic impacts are caused by fouling organisms that require increased or constant cleaning to prevent pipes, hulls and intakes becoming heavily encrusted. For example, the Carpet Sea Squirt (*Didemnum vexillum*) has been predicted to cost Scottish ports and harbours up to £5 million a year if left untreated. Its effects on aquaculture and other marine-based industries are just as severe with potential eradication costs also running into the millions (see Hambrey Consulting, 2011).

The Carpet Sea Squirt is expected to reach the Channel Islands in the short to medium term, probably entering via a harbour or marina. If it can be identified quickly then eradication and management might be possible at a minimal cost. With other species, such as the Slipper Limpet, the problem may already be too great to be dealt with easily or cheaply.

Once in the wider environment, it can be difficult or impossible to contain a prolific non-native species. Predicting the impact on local ecosystems and species can be gauged by looking at the effect that a non-native species has

*Figure 6.5 - A dense carpet of seaweeds and animals fouling the hull of a boat in a French marina. A majority of this fouling is caused by non-native species.*

had in other areas and some mitigation may be possible. Sometimes trying to solve a problem may actually make it worse, such as cutting back the seaweed *Sargassum* in the 1980s which, via drifting specimens, may only have encouraged its spread.

Knowing which biotopes a non-native species will colonise is important as knowledge of the location and extent of a species' preferred habitats can help predict the rate and direction of its spread. This highlights the value of seashore and subtidal habitat mapping, an activity which should be encouraged on all the islands.

Ultimately we shall often be powerless to prevent the arrival and spread of a non-native species but awareness of their potential effect is important. For this reason those species that scored above 30 in the threat assessment should be subject to further investigating to see how they have affected neighbouring regions and how others have dealt with them. In many instances the drawing up a species' action plan will be necessary while with the most extreme cases, such as the Slipper Limpet, additional research work will be required.

## 6.5 - Conclusion

This report contains a baseline study of non-native marine species that are or might affect the Channel Islands. It also offers an assessment of their possible effect, time of arrival and other basic information. The problems posed by non-native marine species to the Channel Islands are the same as in any other part of the English Channel and being aware and being prepared is essential.

Going forwards government departments and NGOs with in the Channel Islands need to develop cooperative monitoring and reporting structures both between themselves and neighbouring countries. The development of action plans is necessary for some species and additional research will be required for others. All this will need to form part of a broader strategy for addressing the increased threats that non-native species (marine and terrestrial) will present to our native species, habitats, landscapes, health and economies.

# 7 - Recommendations for the Channel Islands

### 1 - Protocols for the rapid detection of non-native species

1.1 - Locations at a high risk from non-native species, such as harbours and marinas, should be subject to regular rapid assessment surveys. These could be undertaken by local NGOs.

1.2 - When pontoons and offshore buoys are retrieved they should be inspected by Department of the Environment or NGOs for non-native species.

1.3 - Encourage or sponsor visiting taxonomic experts who are able identify species that are otherwise under-recorded such as annelids, barnacles, tunicates, phytoplankton and red seaweeds.

1.4 - Monitor published and unpublished information sources to horizon scan for non-native species that have been reported from nearby locations and assess the threat from these.

1.5 - Create awareness campaigns for selected species that present particular threats or about which further information is needed such as *Didemnum vexillum* (see also 2.4 below).

### 2 - Information coordination for non-native species

2.1 - Ensure that all survey and other information is coordinated locally (perhaps through the record centres) and then shared nationally through organisations such as NBN Gateway and the GB Non-native Species Secretariat.

2.2 - Create a dedicated website about Channel Islands non-native species which can disseminate information but also collect records from the public. This could be done via the Jersey Biodiversity Centre or the Guernsey Biological Records Centre.

2.3 - Raise awareness of key non-native species with individuals and groups that have regular contact with the marine environment. This could include local fishermen, divers, aquaculture workers, naturalists, boat owners, etc.

2.4 - Use local and social media to raise awareness about individual species and codes of conduct regarding non-native species.

### 3 - Develop protocols for eradicating/excluding non-native species

3.1 - Governments should continue with existing aquaculture biosecurity arrangements and with organisations such as CEFAS.

3.2 - Do not allow the introduction of new aquaculture species without a full and independent assessment of their potential environmental impact. This should include scenarios regarding future rises in sea temperature. Adopt the precautionary principle.

3.3 - Consider creating a code of conduct for local harbours and marinas regarding such things as the cleaning of boat hulls and infrastructure.

### 4 - Further research

4.1 - The American Slipper Limpet (*Crepidula fornicata*) presents a serious threat to the local fishing economy and all shallow marine environments. Further research is urgently required to assess the scale of the problem and the threat that this species is presenting to local habitats and commercial fishing.

4.2 - Action plans may be required for the following species: *Hemigrapsus sanguineus; Crassostrea gigas; Watersipora subatra; Undaria pinnatifida; Sargassum muticum;* and *Grateloupia subpectinata.* Some of this is probably achievable through NGOs.

4.3 - Casual monitoring should be encouraged for all other species.

4.3 - Conduct a complete review of all non-native species every five to ten years. Use this information to establish strategies for existing species and to prepare for the arrival of new species.

# 8 - Endogenic, Cryptogenic and Exogenic Species

This report has primarily concerned itself with those non-native species that have been transported (usually by human agencies) from remote locations and have subsequently established themselves in the Channel Islands biogeographic region. There are, however, other types of species which are sometimes described as being non-native and which have been reported from the Channel Islands area. These can be classified into three main types:

- *Endogenic species*: whose native natural range has expanded to the north or south bringing them into the Normano-Breton Gulf.

- *Cryptogenic species*: possible non-native species that have been established since historical times and whose origin and date of arrival are unknown.

- *Exogenic species*: species have drifted in to the Normano-Breton Gulf but which cannot reproduced locally.

Although not the primary focus of this report, the plants and animals in these groups are sometimes included in the same category as the non-native species described in Chapters Five and Six. For the sake of differentiation and clarity, a summary of these three groups is provided here.

## 8.1 - Endogenic Species

The geographical ranges of marine species is not fixed and will expand and contract over years, decades and centuries. The range of a species will usually be determined by a series of ecological and oceanographic parameters such as the availability of food/nutrients, sea temperature, suitable substrate/host, etc. These parameters can change on a local and regional scale allowing the range of any dependent species to expand or contract as well.

One of the strongest oceanographic factors in determining the range of individual species in the English Channel is sea temperature. Most organisms are adapted to live and breed within a narrow temperature band which restricts their distribution.

In simplistic terms, sea temperature increases towards the south and decreases to the north. The Normano-Breton Gulf is located on the boundary between colder subArctic seas of the north and the temperate waters of the Mediterranean and Biscay to the south (see Chapter Two). Small shifts in sea temperature permit the inclusion or exclusion of species that are living on the fringe of their thermal tolerance. Since 1980 the sea temperature at St Peter Port, Guernsey, has risen by at least 1.7°C (Guernsey Climate Change Partnership, 2007). This warming trend is predicted to continue throughout the twenty-first century which means that warmer water species currently living in the Bay of Biscay may be able to survive in local waters.

The Channel Islands have good marine biological data from two principal periods: (1) between about 1850 and 1920; and (2) between 1982 and the present day. A comparison of fauna and flora from these two periods reveals that since the First World War the marine species within the islands has undergone some subtle changes in terms of abundance and composition. However, with other factors such as regional overfishing and pollution, it is difficult to ascribe these changes solely to environmental change.

There are, however, several large and obvious species which have been consistently recorded during recent decades but which have no pre-World War I records suggesting that they are modern arrivals to the region. These species are here referred to as being of endogenic origin and in most instances their presence in the Normano-Breton Gulf is due to a northward range expansion, suggesting that rising sea temperature may be the cause. There is as yet no definitive list of endogenic species for the Channel Islands region but some of the more salient examples are discussed below.

## Crustaceans

The only crustacean species known to have expanded into the Gulf in recent years is the large parasitic isopod *Ceratothoa steindachneri* which latches on to and then replaces the tongue of its fish host. Although a southerly species, this isopod appears to have recently moved hosts from Mediterranean fishes to the Lesser Weever Fish (*Echiichthys vipera*) which has a more northerly range. It may be that *C. steindachneri* has expanded its range via switching host, rather than through environmental factors such as elevated sea temperatures. It was first found in local waters in 2009 but has probably been in the Gulf since the mid-1990s, when specimens started to turn up elsewhere in the English Channel.

## Molluscs

The molluscs of the Channel Islands have been historically well-documented which makes it easier to spot species that have arrived or disappeared over time. Among the more obvious species changes in recent decades has been a sudden collapse in the Common Octopus (*Octopus vulgaris*) population which, until the early 1960s, occurred in its millions around the Gulf. Jersey marked the northern limit of the octopuses' breeding range and its disappearance is blamed on the exceptionally cold winter of 1962/63 which is believed to have killed a majority the local population. Since this time octopuses have been very rare and most sightings are of the northern Curled Octopus (*Eledone cirrhosa*) which was first recorded in the Channel Islands in 1975. It is thought that the absence of the Common Octopus allowed the range of the Curled Octopus to spread south across the English Channel. This is the only known example of a northern species spreading south into the Normano-Breton Gulf in recent times.

Other molluscs that have become more common include the southern sea slug *Dendrodoris limbata* a Mediterranean species which was found in northern Brittany in 2000 and off Jersey from 2014. Also becoming more

210

*Ceratothoa steindachneri*
Tongue-eating Isopod

*Aplysia depilans*
Sea Hare

*Octopus vulgaris*
Common Octopus

*Balistes capriscus*
Grey Triggerfish

*Sarda sarda*
Bonito

*Scomberesox saurus*
Atlantic Saury

*Figure 8.1 - A selection of endogenic species from Channel Islands waters.*

common are the Sea Hare *Aplysia depilans* and *A. fasciata* both of which were historically rarely reported but which are now seen most years (sometimes in numbers).

## Fish

The ability of some pelagic fish to migrate long distances makes it easy for them to expand and contract their range, sometimes on a seasonal basis. Fish species will accidentally stray outside of their normal range causing individual animals to turn up in unexpected places (e.g. Hammerhead Sharks in the English Channel). Other fish will have had a once wide natural range diminished by overfishing so that when specimens do turn up it is a noteworthy event (e.g. tuna off the coast of Brittany). In neither of these circumstances is a species considered to be endogenic; this term is reserved for a recently arrived fish species that is afterwards consistently reported from the Gulf indicating that it has become part of their natural range.

A good example of this is the Grey Triggerfish (*Balistes capriscus*), a southern European species which was first recorded off St Malo in 1930. Sporadic records then occurred during the 1950s and 60s but by the 1980s captures off Jersey being a regular event and the triggerfish was described as being common in the 1990s.

This pattern of a southern European fish entering the Normano-Breton Gulf and afterwards becoming regularly reported may be seen with the Bonito (*Sarda sarda*), which has been caught from 2004, Red Porgy (*Pagrus pagrus*; from mid-1990s), White Sargo (*Diplodus sargus*; from 1990s) and Atlantic Saury (*Scomberesox saurus*; from circa 2009). Other southerly fish species, such as the Eagle Ray (*Myliobatis aquila*) and Bogue (*Boops boops*), are also being reported more frequently suggesting that their range may be expanding northwards.


## 8.2 - Cryptogenic Species

There are marine species which, while appearing to be native, have characteristics that are associated with non-native species (Chapter 1.1). Often these species will have a global distribution, making it difficult to know whether they have a cosmopolitan range or whether they were transported around the world centuries ago, prior to the advent of biological recording. Such species are often referred to as cryptogenic meaning 'of unknown origin'. Due to their uncertain status, cryptogenic species have not been included with the non-native marine species listed in the main part of this report. Local examples could include:

*Calyptraea chinensis* – Chinaman's Hat
Native to southern Europe, the northern limit of this mollusc is believed to have been living in the Normano-Breton Gulf until the end of the nineteenth century, when it became established in the North Sea. The earliest Channel Islands record is 1810 but archaeological work on seabed cores between

*Calyptraea chinensis*
Chinaman's Hat

*Phallusia mammillata*
Neptune's Heart

*Ciona intestinalis*
Common Sea Squirt

*Pylaiella littoralis (GJ)*

*Janthina janthina*
Purple Sea Snail

*Velella velella*
By-the-Wind Sailor

*Figure 8.2 - A selection of cryptogenic species (top and middle row) and exogenic species (bottom row) from Channel Islands waters.*

Jersey and France suggests that this species may have entered the Gulf just a few centuries ago, although the exact timing is uncertain (P. Chambers, pers. comm.).

*Phallusia mammillata* (Neptune's Heart) and *Ciona intestinalis* (Common Sea Squirt)
These large ascidians are common on European coasts but their global distribution and recent genetic studies suggests that they may have been transported between continents in recent historical times.

*Schizoporella unicornis*
A species that is believed to be native to Japan but which is now cosmopolitan in temperate seas. It is common on the seashore across the Channel Islands.

*Pylaiella littoralis*
Possibly originating from the North-west Pacific this brown seaweed has become cosmopolitan in many parts of the world. It can be prolific in the spring and summer and will sometimes form 'brown tides'.

*Polysiphonia morrowi*
This red seaweed may originate from the North-west Pacific. *P. morrow* has been known from the Normano-Breton Gulf for a long time but modern genetic work suggests that it may have been reintroduced several times, possibly via aquaculture. It has not been recorded from the Channel Islands but this is a difficult species to identify and so could be present.


## 8.3 - Exogenic Species

The English Channel has prevailing westerly winds and is located at the eastern end of the Gulf Stream which results in plants, animals and debris being swept into the region from the open Atlantic Ocean. Sometimes this debris will have tropical or subtropical species attached to it, such as barnacles, piddocks, etc. More rarely, entire animals, such as turtles and jellyfish, may be washed into the Gulf from the open ocean. The cold seas of the English Channel (and other factors) will prevent these species from being able to reproduce and so, while they may be alive when they arrive in the Normano-Breton Gulf, they will often die soon afterwards.

Although these plants and animals are technically non-native species, they are unlikely ever to establish themselves in local waters and so are not included in the main part of this report. For the sake of completeness, a list of exogenic species known from the Channel Islands is presented below (with the time range of reports in brackets) but without commentary.


### Pelagic Cnidaria
*Velella velella* - By-the-Wind Sailor (1912 to 2017)
*Physalia physalis* - Portuguese Man-of-War (1892 to 2009)

## Crustaceans

*Planes minutus* - Columbus's Crab (*c*. 1865)

## Molluscs

*Janthina janthina* - Purple Sea Snail (*c*. 1865 to 1917)
*Bulla striata* - Atlantic Bubble Shell (1877, 1911)
*Bankia bipennata* - (*c*. 1841)
*Bankia fimbriatula* - (*c*. 1865)
*Lyrodus bipartitus* - (*c*. 1865)
*Spathoteredo patha* - (1860)
*Teredora malleolus* - (*c*. 1841)
*Teredothyra excavata* - (*c*. 1865)
*Spirula spirula* - Ram's Horn (*c*. 1865)

## Marine Reptiles

*Caretta caretta* - Loggerhead Turtle (1954 to 2016)
*Lepidochelys kempi* - Kemp's Ridley Turtle (1938 to 1995)
*Eretmochelys imbricata* - Hawkbill Turtle (1948)
*Dermochelys coriacea* - Leatherback Turtle (1965 to 2012)

## Seaweeds

*Sargassum natans* - Gulf Weed (1908, 1947)

# Appendix I

## - Shortlist of Non-native Species from the Channel Islands Region -

This appendix contains a shortlist of non-native species that have already been recorded from the Channel Islands (denoted by an '*') or which have the potential to reach the islands in the near future.

The list is presented in taxonomic order and provides information on each species' region of origin, the probable transport vector that brought it to Europe and the first year in which it was recorded.

| Name | Phylum | Origin | Vector | Arrival |
|---|---|---|---|---|
| *Marteilia refringens* | Cercozoa | Cryptogenic | Aquaculture | 1970 |
| *Bonamia ostreae** | Cercozoa | NE Pacific | Aquaculture | 1978 |
| *Haplosporidium nelsoni* | Cercozoa | NW Pacific | Aquaculture | 1993 |
| *Alexandrium affine* | Myzozoa | NW Pacific | Shipping | 1987 |
| *Alexandrium leei* | Myzozoa | NW Pacific | Shipping | 1991 |
| *Alexandrium minutum* | Myzozoa | Cryptogenic | Shipping | 1985 |
| *Karenia brevisulcata* | Myzozoa | NW Pacific | Unknown | 2009 |
| *Karenia umbella* | Myzozoa | SW Pacific | Unknown | 2008 |
| *Karenia papilionacea* | Myzozoa | NW Pacific | Shipping | 1994 |
| *Takayama tasmanica* | Myzozoa | SW Pacific | Unknown | 2008 |
| *Celtodoryx ciocalyptoides* | Porifera | NW Pacific | Unknown | 1999 |
| *Nemopsis bachei* | Cnidaria | W Atlantic | Shipping | 1851 |
| *Blackfordia virginica* | Cnidaria | Cryptogenic | Shipping | 1953 |
| *Cordylophora caspia* | Cnidaria | Indo-Pacific | Shipping | 1816 |
| *Gonionemus vertens* | Cnidaria | NW Pacific | Aquaculture | 1913 |
| *Diadumene cincta* | Cnidaria | N Pacific | Shipping | 1925 |
| *Diadumene lineata* | Cnidaria | NW Pacific | Aquaculture | 1896 |
| *Mnemiopsis leidyi* | Ctenophora | W Atlantic | Shipping | 2005 |
| *Pseudodactylogyrus anguillae* | Platyhelminthes | NW Pacific | Unknown | 1977 |
| *Koinostylochus ostreophagus* | Platyhelminthes | NW Pacific | Aquaculture | 1970 |
| *Anguillicoloides crassus* | Nematoda | NW Pacific | Aquaculture | 1982 |
| *Boccardia semibranchiata* | Annelida | Mediterranean | Aquaculture | 1990 |
| *Hydroides dianthus* | Annelida | NW Atlantic | Shipping | 1927 |
| *Hydroides elegans* | Annelida | Indo-Pacific | Shipping | 1937 |
| *Hydroides ezoensis* | Annelida | Pacific | Shipping | 1978 |
| *Desdemona ornata* | Annelida | Indo-Pacific | Unknown | 1977 |
| *Ficopomatus enigmaticus* | Annelida | SW Pacific | Aquaculture | 1921 |
| *Neodexiospira brasiliensis** | Annelida | SW Atlantic | Shipping | 1974 |
| *Pileolaria berkeleyana* | Annelida | NW Pacific | Shipping | 1974 |
| *Ammothea hilgendorfi* | Arthropoda | NW Atlantic | Shipping | 1976 |
| *Acartia (Acanthacartia) tonsa* | Arthropoda | Cryptogenic | Shipping | 1927 |
| *Mytilicola intestinalis* | Arthropoda | Mediterranean | Aquaculture | 1938 |
| *Mytilicola orientalis* | Arthropoda | NW Pacific | Aquaculture | 1977 |
| *Pseudomyicola spinosus* | Arthropoda | Pacific | Aquaculture | 1963 |
| *Myicola ostreae* | Arthropoda | NW Pacific | Aquaculture | 1972 |
| *Austrominius modestus** | Arthropoda | S Pacific | Shipping | 1945 |

| | | | | |
|---|---|---|---|---|
| *Hesperibalanus fallax* | Arthropoda | NW Pacific | Shipping | 1994 |
| *Amphibalanus amphitrite* | Arthropoda | SW Pacific | Shipping | 1914 |
| *Amphibalanus eburneus* | Arthropoda | W Atlantic | Shipping | 1954 |
| *Amphibalanus improvisus\** | Arthropoda | W Atlantic | Shipping | 1854 |
| *Amphibalanus reticulatus* | Arthropoda | NW Pacific | Shipping | 1997 |
| *Balanus trigonus* | Arthropoda | Indo-Pacific | Shipping | 1997 |
| *Amphibalanus variegatus* | Arthropoda | S Pacific | Shipping | 1997 |
| *Megabalanus coccopoma* | Arthropoda | S Pacific | Shipping | 1851 |
| *Megabalanus tintinnabulum* | Arthropoda | Indo-Pacific | Shipping | 1764 |
| *Eusarsiella zostericola* | Arthropoda | W Atlantic | Aquaculture | 1870 |
| *Odontodactylus scyllarus* | Arthropoda | Indo-Pacific | Aquarium | 2009 |
| *Grandidierella japonica* | Arthropoda | NW Pacific | Shipping | 1997 |
| *Monocorophium sextonae\** | Arthropoda | SW Pacific | Shipping | 1934 |
| *Caprella mutica* | Arthropoda | NW Pacific | Shipping | 1993 |
| *Limnoria quadripunctata* | Arthropoda | Indo-Pacific | Aquaculture | 1949 |
| *Limnoria tripunctata* | Arthropoda | Indo-Pacific | Aquaculture | 1950 |
| *Penaeus japonicus* | Arthropoda | Indo-Pacific | Aquaculture | 1980 |
| *Palaemon macrodactylus* | Arthropoda | NW Pacific | Shipping | 1992 |
| *Homarus americanus* | Arthropoda | NW Atlantic | Deliberate | 1988 |
| *Asthenognathus atlanticus* | Arthropoda | Mediterranean | Unknown | 2008 |
| *Rhithropanopeus harrisii* | Arthropoda | NW Atlantic | Shipping | 1874 |
| *Hemigrapsus sanguineus\** | Arthropoda | NW Pacific | Shipping | 1999 |
| *Hemigrapsus takanoi* | Arthropoda | NW Pacific | Aquaculture | 1993 |
| *Pachygrapsus marmoratus* | Arthropoda | Mediterranean | Shipping | 1996 |
| *Gibbula albida* | Mollusca | Mediterranean | Aquaculture | 1986 |
| *Potamopyrgus antipodarum\** | Mollusca | SW Pacific | Unknown | 1859 |
| *Crepidula fornicata\** | Mollusca | NW Atlantic | Shipping | 1872 |
| *Tritia neritea* | Mollusca | Mediterranean | Aquaculture | 1984 |
| *Fusinus rostratus* | Mollusca | Mediterranean | Aquaculture | 2007 |
| *Ocenebra inornata* | Mollusca | NW Pacific | Aquaculture | 1995 |
| *Rapana venosa* | Mollusca | NW Pacific | Deliberate | 1950 |
| *Urosalpinx cinerea\** | Mollusca | NW Atlantic | Aquaculture | 1927 |
| *Harmioea japonica* | Mollusca | NW Pacific | Aquaculture | 1992 |
| *Crassostrea gigas\** | Mollusca | NW Pacific | Aquaculture | 1964 |
| *Choromytilus chorus* | Mollusca | SW Atlantic | Aquaculture | 1967 |
| *Mizuhopecten yessoensis* | Mollusca | NW Pacific | Deliberate | 1977 |
| *Ensis directus* | Mollusca | NW Atlantic | Shipping | 1978 |
| *Mercenaria mercenaria\** | Mollusca | NW Atlantic | Deliberate | 1861 |
| *Ruditapes philippinarum\** | Mollusca | NW Pacific | Deliberate | 1970 |
| *Mya arenaria\** | Mollusca | NW Atlantic | Deliberate | 1245 |
| *Mytilopsis leucophaeata* | Mollusca | W Atlantic | Shipping | 1835 |
| *Lyrodus pedicellatus\** | Mollusca | Cryptogenic | Aquaculture | 1849 |
| *Teredo navalis\** | Mollusca | W Atlantic | Aquaculture | 1516 |
| *Victorella pavida* | Bryozoa | Cryptogenic | Shipping | 1960 |
| *Bugula neritina\** | Bryozoa | SW Pacific | Shipping | 1904 |
| *Bugulina stolonifera\** | Bryozoa | NW Pacific | Shipping | 1875 |
| *Caulibugula zanzibariensis* | Bryozoa | Indo-Pacific | Shipping | 2003 |
| *Tricellaria inopinata* | Bryozoa | Indo-Pacific | Unknown | 1998 |
| *Watersipora subatra\** | Bryozoa | NW Pacific | Aquaculture | 1983 |
| *Schizoporella errata* | Bryozoa | Mediterranean | Shipping | 1970 |

217

| | | | | |
|---|---|---|---|---|
| *Schizoporella japonica* | Bryozoa | NW Pacific | Shipping | 2013 |
| *Didemnum vexillum* | Chordata | NW Pacific | Aquaculture | 1998 |
| *Perophora japonica\** | Chordata | NW Pacific | Aquaculture | 1982 |
| *Goniadella gracilis* | Annelida | NW Atlantic | Shipping | 1970 |
| *Corella eumyota\** | Chordata | S Pacific | Unknown | 2004 |
| *Styela clava\** | Chordata | NW Pacific | Aquaculture | 1953 |
| *Asterocarpa humilis* | Chordata | S Pacific | Aquaculture? | 2005 |
| *Botrylloides diegensis\** | Chordata | NW Pacific | Aquaculture? | 2002 |
| *Botrylloides violaceus\** | Chordata | NW Pacific | Aquaculture | 2000 |
| *Molgula manhattensis* | Chordata | W Atlantic | Shipping | 1762 |
| *Oncorhynchus kisutch* | Chordata | NE Pacific | Aquaculture | 1800 |
| *Fibrocapsa japonica* | Ochrophyta | NW Pacific | Shipping | 1991 |
| *Heterosigma akashiwo* | Ochrophyta | NW Pacific | Unknown | 1977 |
| *Pseudo-nitzschia multistriata* | Ochrophyta | NW Pacific | Shipping | 1970s |
| *Stephanopyxis palmeriana* | Ochrophyta | Pacific | Shipping | 1954 |
| *Thalassiosira punctigera* | Ochrophyta | Pacific | Shipping | 1978 |
| *Thalassiosira tealata* | Ochrophyta | Pacific | Shipping | 1995 |
| *Corethron pennatum\** | Ochrophyta | Pacific | Shipping | 1954 |
| *Coscinodiscus wailesii\** | Ochrophyta | Pacific | Shipping | 1977 |
| *Odontella sinensis\** | Ochrophyta | Pacific | Shipping | 1903 |
| *Pleurosigma simonsenii* | Ochrophyta | NE Pacific | aquaculture | 1974 |
| *Colpomenia peregrina\** | Ochrophyta | NE Pacific | Aquaculture | 1905 |
| *Undaria pinnatifida\** | Ochrophyta | NW Pacific | Aquaculture | 1971 |
| *Sargassum muticum\** | Ochrophyta | NW Pacific | Aquaculture | 1971 |
| *Asparagopsis armata\** | Rhodophyta | SW Pacific | Aquaculture | 1925 |
| *Bonnemaisonia hamifera\** | Rhodophyta | NW Pacific | Shipping | 1890 |
| *Grateloupia subpectinata\** | Rhodophyta | NW Pacific | Aquaculture | 1974 |
| *Grateloupia turuturu\** | Rhodophyta | NW Pacific | Aquaculture | 1969 |
| *Polyopes lancifolius\** | Rhodophyta | NW Pacific | Aquaculture | 2008 |
| *Caulacanthus ustulatus* | Rhodophyta | NW Pacific | Aquaculture | 1986 |
| *Pikea californica* | Rhodophyta | Pacific | Shipping | 1990 |
| *Sarcodiotheca gaudichaudii* | Rhodophyta | Pacific | Shipping | 1966 |
| *Solieria chordalis\** | Rhodophyta | Mediterranean | Shipping | 1964 |
| *Gracilaria vermiculophylla\** | Rhodophyta | Indo-Pacific | Aquaculture | 2005 |
| *Lomentaria hakodatensis* | Rhodophyta | NW Pacific | Aquaculture | 1984 |
| *Aglaothamnion halliae* | Rhodophyta | NW Pacific | Shipping | 2004 |
| *Anotrichium furcellatum* | Rhodophyta | N Pacific | Shipping | 1922 |
| *Antithamnion densum* | Rhodophyta | SW Atlantic | Unknown | 1964 |
| *Antithamnion nipponicum* | Rhodophyta | NW Pacific | Shipping | 2007 |
| *Antithamnion pectinatum* | Rhodophyta | NW Pacific | Shipping | 1988 |
| *Antithamnionella spirographidis* | Rhodophyta | N Pacific | Shipping | 1906 |
| *Antithamnionella ternifolia* | Rhodophyta | S Pacific | Shipping | 1906 |
| *Spongoclonium caribaeum* | Rhodophyta | Cryptogenic | Unknown | 1960 |
| *Dasysiphonia japonica* | Rhodophyta | NW Pacific | Aquaculture | 1984 |
| *Laurencia brongniartii* | Rhodophyta | NW Pacific | Aquaculture | 1989 |
| *Neosiphonia harveyi* | Rhodophyta | NW Pacific | Aquaculture | 1908 |
| *Caulerpa taxifolia* | Chlorophyta | Indo-Pacific | Deliberate | 1984 |
| *Codium fragile fragile\** | Chlorophyta | NW Pacific | Unknown | 1940s? |

# Appendix II

## - Threat Assessment Scores for Non-native Species -

This appendix contains the threat assessment scores for each of the non-native species listed in Appendix I. The list is ranked by the overall threat score, highest to lowest. One represents the least threat; five the greatest (see Chapter 2.4 for more details).

**Threat Score** = Overall threat score ; **Ecol** = Impact on ecosystem functions; **Hab** = Impact on native habitats and species; **Tox** = The health impact of a species through disease, toxins, etc.; Econ = The economic impact of a species; **CI** = Dispersal within the Channel Islands; **Hor** = Horizon scanning

| Name | Threat Score | Ecol | Hab | Tox | Econ | CI | Hor |
|---|---|---|---|---|---|---|---|
| Crepidula fornicata | 125 | 5 | 5 | 1 | 5 | 5 | 1 |
| Sargassum muticum | 100 | 5 | 5 | 1 | 4 | 4 | 1 |
| Rapana venosa | 80 | 2 | 5 | 2 | 4 | 1 | 2 |
| Didemnum vexillum | 80 | 4 | 4 | 1 | 5 | 1 | 5 |
| Crassostrea gigas | 72 | 3 | 4 | 3 | 2 | 4 | 1 |
| Pseudodactylogyrus anguillae | 64 | 2 | 4 | 4 | 2 | 1 | 1 |
| Mnemiopsis leidyi | 60 | 3 | 4 | 1 | 5 | 1 | 4 |
| Heterosigma akashiwo | 48 | 2 | 2 | 4 | 3 | 3 | 1 |
| Schizoporella japonica | 36 | 3 | 4 | 1 | 3 | 1 | 3 |
| Undaria pinnatifida | 36 | 3 | 4 | 1 | 3 | 5 | 1 |
| Homarus americanus | 30 | 2 | 5 | 1 | 3 | 1 | 4 |
| Hemigrapsus sanguineus | 30 | 3 | 5 | 1 | 2 | 5 | 1 |
| Pachygrapsus marmoratus | 30 | 3 | 5 | 1 | 2 | 1 | 3 |
| Asterocarpa humilis | 27 | 3 | 3 | 1 | 3 | 1 | 4 |
| Megabalanus coccopoma | 27 | 3 | 3 | 1 | 3 | 1 | 2 |
| Megabalanus tintinnabulum | 27 | 3 | 3 | 1 | 3 | 1 | 2 |
| Bugula neritina | 27 | 3 | 3 | 1 | 3 | 4 | 1 |
| Bugulina stolonifera | 27 | 3 | 3 | 1 | 3 | 4 | 1 |
| Schizoporella errata | 27 | 3 | 3 | 1 | 3 | 1 | 2 |
| Corella eumyota | 27 | 3 | 3 | 1 | 3 | 4 | 1 |
| Styela clava | 27 | 3 | 3 | 1 | 3 | 4 | 1 |
| Alexandrium minutum | 24 | 2 | 2 | 3 | 2 | 3 | 1 |
| Haminoea japonica | 24 | 2 | 2 | 3 | 2 | 1 | 3 |
| Koinostylochus ostreophagus | 24 | 2 | 2 | 3 | 2 | 1 | 3 |
| Mytilicola orientalis | 24 | 2 | 2 | 3 | 2 | 3 | 1 |
| Pseudomyicola spinosus | 24 | 2 | 2 | 3 | 2 | 3 | 1 |
| Hemigrapsus takanoi | 24 | 3 | 4 | 1 | 2 | 1 | 4 |
| Ocenebra inornata | 24 | 2 | 3 | 1 | 4 | 1 | 3 |
| Watersipora subatra | 24 | 3 | 4 | 1 | 2 | 5 | 1 |
| Asparagopsis armata | 24 | 3 | 4 | 1 | 2 | 5 | 1 |
| Grateloupia subpectinata | 24 | 3 | 4 | 1 | 2 | 5 | 1 |
| Grateloupia turuturu | 24 | 3 | 4 | 1 | 2 | 5 | 1 |
| Bonamia ostreae | 20 | 1 | 1 | 4 | 5 | 4 | 1 |
| Celtodoryx ciocalyptoides | 20 | 5 | 4 | 1 | 1 | 1 | 4 |

| | | | | | | |
|---|---|---|---|---|---|---|
| Tricellaria inopinata | 18 | 2 | 3 | 1 | 3 | 4 | 1 |
| Perophora japonica | 18 | 3 | 3 | 1 | 2 | 4 | 1 |
| Molgula manhattensis | 18 | 3 | 3 | 1 | 2 | 3 | 1 |
| Caulacanthus ustulatus | 18 | 3 | 3 | 1 | 2 | 1 | 4 |
| Marteilia refringens | 16 | 1 | 1 | 4 | 4 | 1 | 2 |
| Ammothea hilgendorfi | 16 | 4 | 4 | 1 | 1 | 1 | 5 |
| Mytilicola intestinalis | 16 | 2 | 2 | 1 | 4 | 1 | 2 |
| Myicola ostreae | 16 | 2 | 2 | 2 | 2 | 1 | 2 |
| Dasysiphonia japonica | 16 | 4 | 4 | 1 | 1 | 4 | 1 |
| Anguillicoloides crassus | 15 | 1 | 5 | 3 | 1 | 3 | 1 |
| Gonionemus vertens | 12 | 2 | 2 | 3 | 1 | 3 | 1 |
| Desdemona ornata | 12 | 3 | 4 | 1 | 1 | 1 | 2 |
| Grandidierella japonica | 12 | 3 | 4 | 1 | 1 | 1 | 2 |
| Limnoria quadripunctata | 12 | 2 | 2 | 1 | 3 | 1 | 2 |
| Limnoria tripunctata | 12 | 2 | 2 | 1 | 3 | 1 | 2 |
| Ensis directus | 12 | 3 | 4 | 1 | 1 | 1 | 2 |
| Coscinodiscus wailesii | 12 | 2 | 3 | 1 | 2 | 4 | 1 |
| Solieria chordalis | 12 | 2 | 3 | 1 | 2 | 5 | 1 |
| Caulerpa taxifolia | 12 | 3 | 4 | 1 | 1 | 1 | 2 |
| Codium fragile fragile | 12 | 2 | 3 | 1 | 2 | 4 | 1 |
| Haplosporidium nelsoni | 9 | 1 | 1 | 3 | 3 | 1 | 2 |
| Alexandrium affine | 9 | 1 | 1 | 3 | 3 | 3 | 1 |
| Alexandrium leei | 9 | 1 | 1 | 3 | 3 | 3 | 1 |
| Karenia brevisulcata | 9 | 1 | 1 | 3 | 3 | 3 | 1 |
| Fusinus rostratus | 9 | 3 | 3 | 1 | 1 | 1 | 4 |
| Antithamnion pectinatum | 9 | 3 | 3 | 1 | 1 | 1 | 2 |
| Karenia umbella | 8 | 1 | 1 | 4 | 2 | 1 | 2 |
| Karenia papilionacea | 8 | 1 | 1 | 4 | 2 | 1 | 2 |
| Takayama tasmanica | 8 | 1 | 1 | 4 | 2 | 1 | 3 |
| Asthenognathus atlanticus | 8 | 2 | 2 | 1 | 2 | 1 | 3 |
| Hesperibalanus fallax | 8 | 2 | 2 | 1 | 2 | 4 | 1 |
| Amphibalanus reticulatus | 8 | 2 | 2 | 1 | 2 | 1 | 2 |
| Balanus trigonus | 8 | 2 | 2 | 1 | 2 | 1 | 2 |
| Amphibalanus variegatus | 8 | 2 | 2 | 1 | 2 | 1 | 2 |
| Eusarsiella zostericola | 8 | 2 | 2 | 1 | 2 | 1 | 2 |
| Penaeus japonicus | 8 | 2 | 4 | 1 | 1 | 1 | 2 |
| Urosalpinx cinerea | 8 | 1 | 1 | 2 | 4 | 2 | 1 |
| Botrylloides violaceus/B.diegensis | 8 | 2 | 4 | 1 | 1 | 4 | 1 |
| Antithamnionella spirographidis | 8 | 2 | 2 | 1 | 2 | 3 | 1 |
| Neosiphonia harveyi | 8 | 2 | 2 | 1 | 2 | 4 | 1 |
| Amphibalanus amphitrite | 6 | 2 | 3 | 1 | 1 | 3 | 1 |
| Caprella mutica | 6 | 2 | 3 | 1 | 1 | 1 | 3 |
| Palaemon macrodactylus | 6 | 2 | 3 | 1 | 1 | 1 | 1 |
| Gibbula albida | 6 | 2 | 3 | 1 | 1 | 1 | 2 |
| Ruditapes philippinarum | 6 | 2 | 3 | 1 | 1 | 4 | 1 |
| Gracilaria vermiculophylla | 6 | 2 | 3 | 1 | 1 | 4 | 1 |
| Aglaothamnion halliae | 6 | 2 | 3 | 1 | 1 | 1 | 4 |
| Nemopsis bachei | 4 | 2 | 2 | 1 | 1 | 3 | 1 |
| Blackfordia virginica | 4 | 2 | 2 | 1 | 1 | 1 | 1 |
| Cordylophora caspia | 4 | 2 | 2 | 1 | 1 | 1 | 1 |

| | | | | | | |
|---|---|---|---|---|---|---|
| Diadumene cincta | 4 | 2 | 2 | 1 | 1 | 1 | 3 |
| Diadumene lineata | 4 | 2 | 2 | 1 | 1 | 4 | 1 |
| Hydroides dianthus | 4 | 2 | 2 | 1 | 1 | 3 | 1 |
| Hydroides elegans | 4 | 2 | 2 | 1 | 1 | 3 | 1 |
| Hydroides ezoensis | 4 | 2 | 2 | 1 | 1 | 3 | 1 |
| Ficopomatus enigmaticus | 4 | 1 | 2 | 1 | 2 | 1 | 2 |
| Neodexiospira brasiliensis | 4 | 2 | 2 | 1 | 1 | 1 | 3 |
| Pileolaria berkeleyana | 4 | 2 | 2 | 1 | 1 | 4 | 1 |
| Acartia (Acanthacartia) tonsa | 4 | 2 | 2 | 1 | 1 | 1 | 1 |
| Austrominius modestus | 4 | 2 | 2 | 1 | 1 | 4 | 1 |
| Amphibalanus eburneus | 4 | 2 | 2 | 1 | 1 | 1 | 4 |
| Amphibalanus improvisus | 4 | 2 | 2 | 1 | 1 | 2 | 1 |
| Odontodactylus scyllarus | 4 | 2 | 2 | 1 | 1 | 1 | 1 |
| Monocorophium sextonae | 4 | 2 | 2 | 1 | 1 | 4 | 1 |
| Potamopyrgus antipodarum | 4 | 2 | 2 | 1 | 1 | 4 | 1 |
| Mercenaria mercenaria | 4 | 2 | 2 | 1 | 1 | 2 | 1 |
| Caulibugula zanzibariensis | 4 | 2 | 2 | 1 | 1 | 1 | 2 |
| Goniadella gracilis | 4 | 2 | 2 | 1 | 1 | 1 | 2 |
| Fibrocapsa japonica | 4 | 1 | 1 | 2 | 2 | 3 | 1 |
| Colpomenia peregrina | 4 | 2 | 2 | 1 | 1 | 4 | 1 |
| Bonnemaisonia hamifera | 4 | 2 | 2 | 1 | 1 | 4 | 1 |
| Polyopes lancifolius | 4 | 2 | 2 | 1 | 1 | 4 | 1 |
| Sarcodiotheca gaudichaudii | 4 | 2 | 2 | 1 | 1 | 1 | 1 |
| Lomentaria hakodatensis | 4 | 2 | 2 | 1 | 1 | 1 | 3 |
| Anotrichium furcellatum | 4 | 2 | 2 | 1 | 1 | 3 | 1 |
| Antithamnion densum | 4 | 2 | 2 | 1 | 1 | 3 | 1 |
| Antithamnion nipponicum | 4 | 2 | 2 | 1 | 1 | 1 | 2 |
| Antithamnionella ternifolia | 4 | 2 | 2 | 1 | 1 | 4 | 1 |
| Spongoclonium carlbaeum | 4 | 2 | 2 | 1 | 1 | 3 | 1 |
| Laurencia brongniartii | 4 | 2 | 2 | 1 | 1 | 3 | 1 |
| Lyrodus pedicellatus | 3 | 1 | 1 | 1 | 3 | 4 | 1 |
| Teredo navalis | 3 | 1 | 1 | 1 | 3 | 4 | 1 |
| Boccardia semibranchiata | 2 | 2 | 1 | 1 | 1 | 1 | 2 |
| Mya arenaria | 2 | 2 | 1 | 1 | 1 | 1 | 1 |
| Oncorhynchus kisutch | 2 | 1 | 2 | 1 | 1 | 1 | 1 |
| Thalassiosira tealata | 2 | 2 | 1 | 1 | 1 | 1 | 5 |
| Odontella sinensis | 2 | 1 | 2 | 1 | 1 | 4 | 1 |
| Rhithropanopeus harrisii | 1 | 1 | 1 | 1 | 1 | 1 | 1 |
| Tritia neritea | 1 | 1 | 1 | 1 | 1 | 1 | 1 |
| Choromytilus chorus | 1 | 1 | 1 | 1 | 1 | 1 | 1 |
| Mizuhopecten yessoensis | 1 | 1 | 1 | 1 | 1 | 1 | 1 |
| Mytilopsis leucophaeata | 1 | 1 | 1 | 1 | 1 | 1 | 1 |
| Victorella pavida | 1 | 1 | 1 | 1 | 1 | 1 | 1 |
| Pseudo-nitzschia multistriata | 1 | 1 | 1 | 1 | 1 | 3 | 1 |
| Stephanopyxis palmeriana | 1 | 1 | 1 | 1 | 1 | 3 | 1 |
| Thalassiosira punctigera | 1 | 1 | 1 | 1 | 1 | 3 | 1 |
| Corethron pennatum | 1 | 1 | 1 | 1 | 1 | 3 | 1 |
| Pleurosigma simonsenii | 1 | 1 | 1 | 1 | 1 | 1 | 5 |
| Pikea californica | 1 | 1 | 1 | 1 | 1 | 1 | 5 |

# Appendix III
## - Channel Island Non-native Marine Species -

This appendix contains a summary of the known distriubtion of non-native marine species within the Channel Islands listed in alphabetical order. It is based on several data sources but principally uses records held by the Société Jersiaise, Jersey Biodiversity Centre and Guernsey Biological Records Centre (see Chapter 2 for more details).

This probably does not represent each species' true distribution as some islands have been better surveyed than others.
Je = Jersey; Ecr = Les Écréhous; Min = Les Minquiers; Pat = Paternosters; Gu = Guernsey; Al = Alderney; Sk = sark; He = Herm; Li = Lihou.

| Scientific Name | Je | Ecr | Min | Pat | Gu | Al | Sk | He | Li |
|---|---|---|---|---|---|---|---|---|---|
| Amphibalanus improvisus | | | | | | | | | |
| Antithamnionella ternifolia | X | | | | X | | | X | |
| Asparagopsis armata | X | | | | X | X | | | |
| Austrominius modestus | X | X | X | | X | X | X | X | X |
| Bonamia ostreae | X | | X | | X | X | | X | |
| Bonnemaisonia hamifera | X | | | | X | X | | | |
| Botrylloides diegensis | X | | | | X | | | | |
| Botrylloides violaceus | X | | | | X | | | | |
| Bugula neritina | X | | | | X | | | | |
| Bugulina stolonifera | X | | | | | | | | |
| Codium fragile fragile | X | | | X | X | X | | | |
| Colpomenia peregrina | X | X | X | | X | X | X | X | |
| Corella eumyota | X | | | | X | | | | |
| Corethron pennatum | X | | | | | | | | |
| Coscinodiscus wailesii | X | | | | | | | | |
| Crassostrea gigas | X | X | X | | X | X | X | | |
| Crepidula fornicata | X | X | X | | X | X | X | | X |
| Dasysiphonia japonica | X | | | | | | | | |
| Diadumene lineata | | | | | X | | | | |
| Gracilaria vermiculophylla | X | | | | | | | | |
| Grateloupia subpectinata | X | X | X | | X | X | | | |
| Grateloupia turuturu | X | | X | | | | | | |
| Hemigrapsus sanguineus | X | X | | | X | | | | |
| Hesperibalanus fallax | | | | | X | | | | |
| Lyrodus pedicellatus | X | | | | X | X | | X | |

| | | | | | | | | |
|---|---|---|---|---|---|---|---|---|
| Mercenaria mercenaria | X | | | | | | | | |
| Monocorophium sextonae | X | | | | | | | | |
| Mya arenaria | | | | | X | | | | |
| Neosiphonia harveyi | X | | | | | | | | |
| Odontella sinensis | X | | | | X | | | | |
| Oncorhynchus kisutch | | | | | X | | | | |
| Perophora japonica | X | | | | X | | | | |
| Pileolaria berkeleyana | X | | | | | | | | |
| Polyopes lancifolius | X | | | | | | | | |
| Potamopyrgus antipodarum | X | | | | X | | | | |
| Ruditapes philippinarum | X | X | | | | | | | |
| Sargassum muticum | X | X | X | X | X | X | X | X | X |
| Solieria chordalis | X | | | | | | | | |
| Styela clava | X | X | X | | X | | | X | |
| Teredo navalis | X | | | | X | X | | | |
| Tricellaria inopinata | X | | | | | | | | |
| Undaria pinnatifida | X | | | | X | | | | |
| Urosalpinx cinerea | X | | | | | | | | |
| Watersipora subatra | X | | | | X | | | | |

The species listed below have biological, geographical or other properties which suggest that they may already be established in the Channel Islands but have yet to be identified and recorded.

*Alexandrium affine*  
*Alexandrium leei*  
*Alexandrium minutum*  
*Karenia brevisulcata*  
*Gonionemus vertens*  
*Nemopsis bachei*  
*Anguillicoloides crassus*  
*Hydroides dianthus*  
*Hydroides elegans*  
*Hydroides ezoensis*  
*Pseudomyicola spinosus*  
*Amphibalanus amphitrite*  
*Mytilicola orientalis*  

*Molgula manhattensis*  
*Fibrocapsa japonica*  
*Heterosigma akashiwo*  
*Pseudo-nitzschia multistriata*  
*Stephanopyxis palmeriana*  
*Thalassiosira punctigera*  
*Corethron pennatum*  
*Anotrichium furcellatum*  
*Antithamnion densum*  
*Antithamnionella spirographidis*  
*Laurencia brongniartii*  
*Spongoclonium caribaeum*

# Appendix IV
## - Non-native Species and Biotope Preference -

Seashore and shallow marine surveys made by Seasearch and the Société Jersiaise link species reports with the biotope in which they were seen. They also afford most species reports with an abundance rating on the SCAFOR scale. The databases of Seasearch and the Société Jersiaise contain 1,321 species records which are linked to individual biotopes as classified according to the JNCC/EUNIS scheme.

These biotopes are listed below together with the non-native species that have been reported from them and their average abundance. The species are listed in order of highest abundance using the following abbreviations: A = abunadnt; C = common; F = frequent; O = occasional; R = rare; P = present.

**LR.HLR.MusB.Cht; A1.112**
**Chthamalus spp. on exposed eulittoral rock**
*Codium fragile fragile* - P; *Crassostrea gigas* - P

**LR.HLR.MusB.Sem; A1.113**
**Semibalanus balanoides on exposed to moderately exposed or vertical sheltered eulittoral rock**
*Polysiphonia harveyi* – O ; *Styela clava* – O ; *Codium fragile fragile* – R; *Crassostrea gigas* – R; *Crepidula fornicata* – R; *Grateloupia subpectinata* – R; *Undaria pinnatifida* - P

**LR.HLR.FR.Coff; A1.122**
**Corallina officinalis on exposed to moderately exposed lower eulittoral rock**
*Sargassum muticum* – O; *Codium fragile fragile* – P; *Crepidula fornicata* - P

**LR.HLR.FR.Mas; A1.125**
**Mastocarpus stellatus and Chondrus crispus on very exposed to moderately exposed lower eulittoral rock**
*Undaria pinnatifida* - O; *Grateloupia subpectinata* - O; *Watersipora subatra* - O; *Crepidula fornicata*   R; *Sargassum muticum* - P; *Styela clava* - P; *Crassostrea gigas* - P; *Solieria chordalis* - P

**LR.HLR.FR.Osm; A1.126**
**Osmundea pinnatifida on moderately exposed mid eulittoral rock**
*Crepidula fornicata* - R

**LR.MLR.BF.Fser; A1.214**
**Fucus serratus on moderately exposed lower eulittoral rock**
*Crepidula fornicata* - O; *Sargassum muticum* - O; *Undaria pinnatifida* - R; *Grateloupia subpectinata* - R; *Sargassum muticum* - R; *Crassostrea gigas* – R; *Watersipora subatra* - P;

**LR.LLR.F.Fspi; A1.312**
**Fucus spiralis on sheltered upper eulittoral rock**
*Crassostrea gigas* - O; *Styela clava* - R; *Codium fragile fragile* - P; *Crepidula fornicata* - P

### LR.LLR.F.Fves ; A1.313
**Fucus vesiculosus on moderately exposed to sheltered mid eulittoral rock**
*Gracilaria vermiculophylla* - F; *Watersipora subatra* - O; *Tapes philippinarum* - O; *Crepidula fornicata* - P; *Crassostrea gigas* - R; *Grateloupia subpectinata* - R; *Sargassum muticum* - R; *Codium fragile fragile* - P; *Botrylloides violaceus* - P

### LR.LLR.F.Asc ; A1.314
**Ascophyllum nodosum on very sheltered mid eulittoral rock**
*Watersipora subatra* - R; *Crassostrea gigas* - R; *Crepidula fornicata* - P; *Sargassum muticum* - P; *Styela clava* - P

### LR.FLR.Rkp.Cor; A1.411
**Coralline crust-dominated shallow eulittoral rockpools**
*Crepidula fornicata* - O; *Codium fragile fragile* - O; *Sargassum muticum* - R

### LR.FLR.Rkp.FK; A1.412
**Fucoids and kelp in deep eulittoral rockpools**
*Grateloupia subpectinata* - O; *Undaria pinnatifida* - O; *Styela clava* - O; *Grateloupia turuturu* - O; *Crepidula fornicata* - R; *Codium fragile fragile* - R; *Sargassum muticum* - R; *Watersipora subatra* - R; *Crassostrea gigas* - P; *Polyopes lancifolius* - P

### LR.FLR.Rkp.SwSed; A1.413
**Seaweeds in sediment-floored eulittoral rockpools**
*Grateloupia turuturu* - O; *Crepidula fornicata* - O; *Grateloupia subpectinata* - R; *Sargassum muticum* - R

### LR.FLR.CvOv.SpR; A1.446
**Sponges and shade-tolerant red seaweeds on overhanging lower eulittoral bedrock and in cave entrances**
*Sargassum muticum* - C; *Crepidula fornicata* - F; *Watersipora subatra* - O; *Styela clava* - P

### LS.LCS.Sh.BarSh; A2.111
**Barren littoral shingle**
*Hemigrapsus sanguineus* - F

### LS.LSa.MoSa.Ol; A2.222
**Oligochaetes in littoral mobile sand**
*Tapes philippinarum* - P

### LS.LSa.FiSa.Po; A2.231
**Polychaetes in littoral fine sand**
*Crepidula fornicata* - R

### LS.LSa.MuSa.MacAre; A2.241
**Macoma balthica and Arenicola marina in littoral muddy sand**
*Gracilaria vermiculophylla* - F; *Crepidula fornicata* - P; *Watersipora subatra* - P; *Styela clava* - P

### LS.LSa.MuSa.CerPo; A2.242
**Cerastoderma edule and polychaetes in littoral muddy sand**
*Crepidula fornicata* - A

### LS.LSa.MuSa.Lan; A2.245
**Lanice conchilega in littoral sand**
*Crepidula fornicata* - O; *Sargassum muticum* - R; *Antithamnionella ternifolia* - R; *Heterosiphonia japonica* - R; *Grateloupia subpectinata* - P

## LS.LMx.Mx.CirCer; A2.421
**Cirratulids and *Cerastoderma edule* in littoral mixed sediment**
*Crepidula fornicata* - C

## LS.LMp.LSgr.Znol; A2.6111
***Zostera noltii* beds in littoral muddy sand**
*Crepidula fornicata* - P; *Sargassum muticum* - P; *Tapes philippinarum* - P

## No JNCC Code; A2.871/A2.872
**Flooded Gully Complexes**
*Asparagopsis armata* - C; *Sargassum muticum* - F; *Undaria pinnatifida* - F; *Crepidula fornicata* - O; *Grateloupia subpectinata* - O; *Watersipora subatra* - O; *Grateloupia turuturu* - O; *Crassostrea gigas* - O; *Botrylloides violaceus* - O; *Polysiphonia harveyi* - O; *Tapes philippinarum* - R; *Styela clava* - R; *Codium fragile fragile* - R; *Elminius modestus* - P

## IR.HIR.KFaR.FoR.Dic ; A3.1161
**Foliose red seaweeds with dense *Dictyota dichotoma* and/or *Dictyopteris membranacea* on exposed lower infralittoral rock**
*Asparagopsis armata* – C; *Crepidula fornicata* - R

## IR.MIR.KR.Ldig; A3.211
***Laminaria digitata* on moderately exposed sublittoral fringe rock**
*Asparagopsis armata* - F; *Sargassum muticum* - O; *Crepidula fornicata* - P

## IR.MIR.KR.XFoR; A3.215
**Dense foliose red seaweeds on silty moderately exposed infralittoral rock**
*Sargassum muticum* - O

## IR.LIR.K.Sar; A3.315
***Sargassum muticum* on shallow slightly tide-swept infralittoral mixed substrata**
*Sargassum muticum* – C; *Asparagopsis armata* - O

## IR.FIR.IFou; A3.72
**Infralittoral fouling seaweed communities**
*Crepidula fornicata* - R

## CR.HCR.XFa.ByErSp; A4.131
**Bryozoan turf and erect sponges on tide-swept circalittoral rock**
*Perophora japonica* – O; *Styela clava* - R

## CR.FCR.FouFa; A4.72
**Circalittoral fouling faunal communities**
*Tricellaria inopinata* - A; *Undaria pinnatifida* - C; *Monocorophium sextonae* - C; *Watersipora subutra* - F; *Crassostrea gigas* - F; *Botrylloides violaceus* - F; *Bugula neritina* - F; *Bugulina stolonifera* - O; *Corella eumyota* - O; *Perophora japonica* - O; *Styela clava* - R; *Sargassum muticum* - P

## SS.SCS.ICS.MoeVen; A5.133
***Moerella* spp. with venerid bivalves in infralittoral gravelly sand**
*Asparagopsis armata* - F; *Perophora japonica* - O; *Sargassum muticum* - O; *Styela clava* - O; *Crepidula fornicata* - O; *Grateloupia subpectinata* - O

## SS.SCS.ICS.SLan; A5.137
**Dense *Lanice conchilega* and other polychaetes in tide-swept infralittoral sand and mixed gravelly sand**
*Crepidula fornicata* - A

### SS.SSa.IFiSa.IMoSa; A5.231
**Infralittoral mobile clean sand with sparse fauna**
*Grateloupia turuturu* – F; *Crepidula fornicata* - O

### SS.SSa.IMuSa; A5.24
**Infralittoral muddy sand**
*Sargassum muticum* - R

### SS.SMx.IMx.CreAsAn; A5.431
***Crepidula fornicata* with ascidians and anenomes on infralittoral coarse mixed sediment**
*Crepidula fornicata* – C; *Sargassum muticum* - R

### SS.SMp.Mrl; A5.51
**Maerl beds**
*Crepidula fornicata* - C; *Asparagopsis armata* - F; *Styela clava* - O; *Sargassum muticum* - P

### SS.SMp.KSwSS; A5.52
**Kelp and seaweed communities on sublittoral sediment**
*Sargassum muticum* – F; *Asparagopsis armata* - O

### SS.SMp.SSgr.Zmar; A5.5331
***Zostera marina/angustifolia* beds on lower shore or infralittoral clean or muddy sand**
*Codium fragile fragile* - C; *Sargassum muticum* - O; *Crepidula fornicata* - O

# References

\* = Recommended for general species identification and information.

Arenas F, Bishop JDD, Carlton JT, Dyrynda PJ, Farnham WF, Gonzalez DJ, Jacobs MW, Lambert C, Lambert G, Nielsen SE, Pederson JA, Porter JS, Ward S, Wood CA, 2006. Alien species and other notable records from a rapid assessment survey of marinas on the south coast of England. *Journal of the Marine Biological Association of the United Kingdom*. Vol. 86: 1329–1337

Ashelby CW, Worsfold TM, Fransen CHJM, 2004. First records of the oriental prawn *Palaemon macrodactylus* (Decapoda: Caridea) an alien species in European waters with a revised key to British Palaemonidae. *Journal of the Marine Biological Association of the United Kingdom*. Vol. 84: 1041– 1050

Ashton G, Boos K, Shucksmith R, Cook E, 2006a. Rapid assessment of the distribution of marine non-native species in marinas in Scotland. *Aquatic Invasions*. Vol. 1: 209–213

Ashton G, Boos K, Shucksmith R, Cook E, 2006b. Risk assessment of hull fouling as a vector for non-natives in Scotland. *Aquatic Invasions*. Vol. 1: 214–218

\*Audibert, C and Delemarre, J-L, 2009. *Guide des coquillages de France: Atlantique et Manche*. Berlin.

Avery JD, Fonseca DM, Campagne P, Lockwood JL, 2013. Cryptic introductions and the interpretation of island biodiversity. *Molecular Ecology*. Vol. 22, 2313–2324.

Bamber RN, 1985. The itinerant sea spider *Ammothea hilgendorfi* (Böhm) in British waters. *Proceedings of Hampshire Field Club & Archaeological Society*. Vol. 41: 269–270

Bamber RN, 1987a. A benthic myodocopid ostracod in Britain. *Porcupine Newsletter*. Vol. 4: 7–9

Bamber RN, 1987b. Some aspects of the biology of the North American ostracod *Sarsiella zostericola* Cushman in the vicinity of a British power station. *Journal of Micropalaeontology*. Vol. 6: 57–62

Bax, N *et al.*, 2003. 'Marine invasive alien species: a threat to global biodiversity,' *Marine Policy*. Vol. 27(4): 313-323.

Bishop, JDD, Roby, C, Yunnie, ALE *et al.*, 2013. The Southern Hemisphere ascidian *Asterocarpa humilis* is unrecognised but widely established in NW France and Great Britain. *Biological Invasions*. Vol. 15: 253.

Blackburn TM, Pysek P, Bacher S *et al.*, 2011. A proposed unified framework for biological invasions. *Trends in Ecology and Evolution*. Vol. 26: 333–339.

Blanchard M, 1995. Origine et état de la population de *Crepidula fornicata* (Gastropoda Prosobranchia) sur le littoral Français. *Haliotus*. Vol. 24: 75-86

Blanchard, M, 1997. Spread of the slipper limpet *Crepidula fornicata* (L. 1758) in Europe. Current state and consequences. *Sci. Mar*. Vol. 61(2): 109-118.

Blanchard, M, 2009. Recent expansion of the slipper limpet population (Crepidula fornicata) in the Bay of Mont-Saint-Michel (Western Channel, France). *Aquatic Living Resources*. Vol. 22: 11-19.

Booy O, White V, Wade M, 2006. *Non-Native organism risk assessment scheme: trialling and peer review.* Scottish Executive reference: FF/05/22.

Bracken, AM, 2012. *A study of the ecology of marine algae in flooded gully habitats on Jersey's south-east coast with an emphasis on the established non-native species* Sargassum muticum *(Yendo) Fensholt.* Unpublished MSc Dissertation, Univeristy of Reading.

Branquart E (ed.), 2007. *Guidelines for environmental impact assessment and list classification of non-native organisms in Belgium.* Belgium Government.

*Brodie, J, Maggs, CA and John, DM, 2007. *Green seaweeds of Britain and Ireland.* British Phycological Society.

Brylinski JM, 1981. Report on the presence of *Acartia tonsa* Dana (Copepoda) in the harbour of Dunkirk (France) and its geographic distribution in Europe. *Journal of Plankton Research.* Vol. 3: 255–260

*Bunker, F, Brodie, JA, Maggs, CA and Bunker, AR, 2017. *Seaweeds of Britain and Ireland.* Wild nature Press.

Carlton JT, Geller JB, 1993. Ecological roulette: The global transport of nonindigenous marine organisms. *Science.* Vol. 261: 78–82

Chambers, P, Binney, F and Jeffreys, G, 2016. *Les Minquiers: a natural history.* Charonia Media.

Cook EJ, Jahnke M, Kerckhof F, Minchin D, Faasse M, Ashton KB, 2007. European expansion of the introduced amphipod *Caprella mutica* Schurin 1935. *Aquatic Invasions.* Vol. 2: 411–421

Critchley AT, Farnham WF, Morrell SL, 1986. An account of the attempted control of an introduced marine alga *Sargassum muticum,* in southern England. *Biological Conservation.* Vol. 35: 313–332

DAISIE, 2009. *Handbook of Alien Species in Europe.* Springer, Dordrecht.

Dansey P, 2011. *Ensis directus* (Conrad 1843) (Bivalvia: Solenoidea) found in Liverpool Bay (Sea area 24). *Journal of Conchology.* Vol. 40: 679

Dauvin J-C, 2009. New record of the marbled crab *Pachygrapsus marmoratus* (Crustacea: Brachyura: Grapsoidea) on the coast of northern Cotentin, Normandy, western English Channel. *Marine Biodiversity Records.* Vol. 2: 1–3

Davis MH, Lutzen J, Davis ME, 2007b. The spread of *Styela clava* Herdman 1882 (Tunicata, ascidiacea) in European waters. *Aquatic Invasions.* Vol. 2: 378–390

De Clercq P, Mason PG, Babendreier D, 2011. Benefits and risks of exotic biological control agents. *BioControl.* Vol. 56: 307–324.

D'Hondt, J-L and Breton, G, 2005. Une nouvelle introduction dans le Bassin d'Arcachon: le bryozoaire cheilostome intertropical *Caulibugula zanzibarensis* waters, 1913. *Science Action (Normandie). Bulletin de la Société Géologique de Normandie.* Vol. 92(1): 19-22.

Dyrynda PEJ, Fairall VR, Occhipinti Ambrogi A, d'Hondt J-L, 2000. The distribution, origins and taxonomy of *Tricellaria inopinata* d'Hondt and Occhipinti Ambrogi, 1985, an invasive bryozoan new to the Atlantic. *Journal of Natural History.* Vol. 34: 1993–2006

Eno CE, 1998. Non-native marine species in British waters: effects and controls. *Aquatic Conservation: Marine and Freshwater Ecosystems.* Vol. 6: 215–228

Eno NC, Clark RA, Sanderson WG, 1997. *Non-native species in British waters: a review and dictionary.* Joint Nature Conservation Committee.

Essl F, Nehring S, Klingenstein F, Milasowszky N, Nowack C, Rabitsch W, 2011. Review of risk assessment systems of IAS in Europe and introducing the German-Austrian Black List Information System (GABLIS). *Journal for Nature Conservation*. Vol. 19: 339–350.

Fletcher RL, Manfredi C, 1995. The occurrence of *Undaria pinnatifida* (Phaeophyta: Laminariales) on the south coast of England. *Botanica Marina*. Vol. 38: 355–358

Foster V, Gieslet RJ, Wilson AMW, Nall CR and Cook, EJ, 2016. Identifying the physical features of marina infrastructure associated with the presence of non-native species in the UK. *Marine Biology*. Vol. 163: e173.

Gallardo B, Aldridge DC, 2013. The 'dirty dozen': socio-economic factors amplify the invasion potential of 12 high-risk aquatic invasive species in Great Britain and Ireland. *Journal of Applied Ecology*. Vol. 50: 757–766.

Godet L, Le Mao P, Grant C and Olivier F, 2010. Marine invertebrate fauna of the Chausey Archipelago: an annotated checklist of historical data from 1828 to 2008. *Cahiers de Biologie Marine*. Vol. 51: 147-165

Gómez F, 2008. Phytoplankton invasions: comments on the validity of categorising the non-indigenous dinoflagellates and diatoms in Eurpean seas. *Marine Pollution Bulletin*. Vol. 56: 620–628

*Goulletquer, P, 2016. *Guide des organismes exotiques marins*. Belin.

Greenway, B, 2001. *Beach and oceanographic processes surrounding Jersey, Channel Islands*. Unpublished MPhil Thesis: University of Southampton.

Hambrey Consulting, 2011. *Cost benefit analysis of management options for Didemnum vexillum in Scotland*. Unpublished tender report for the Scottish Government.

Holt RHF, Ramsay K, Mowat S, Kent FEA, Griffith K, 2008. *Survey of a non-native ascidian (sea-squirt) Didemnum vexillum in Holyhead Marina*. CCW Marine Monitoring Report No: 67.

Holm, NA, 1966. The bottom fauna of the English Channel, part II. *Journal of the Marine Biological Association of the United Kingdom*. Vol. 46: 401-493.

Hulme PE, 2009. Trade, transport and trouble: managing invasive species pathways in an era of globalization. *Journal of Applied Ecology*. Vol. 46: 10–18

Hülsmann N, Galil BS, 2002. Protists – a dominant component of the ballast – transported biota. In: Leppäkoski E, Gollasch S, Olenin S (eds), *Invasive Aquatic Species of Europe: Distribution, Impact and Management*. Dordrecht, Kluwer Academic Publishing, pp 20–26

Ingle RW, Clark PF, 2008. First reported occurrences of the marbled crab, *Pachygrapsus marmoratus* (Crustacea: Brachyura: Grapsoidea) in southern coastal waters of the British Isles. *Marine Biodiversity Records*. Vol. 1: e26

Jensen A, Humphreys J, Caldow R, Grisley C, Dyrynda PEJ, 2004. Naturalisation of the manila clam (*Ruditapes philippinarum*), an alien species and establishment of a clam fishery within Poole Harbour, Dorset. *Journal of the Marine Biological Association of the United Kingdom*. Vol. 84: 1069–1073

Jourde, J, Alizier, S, Dancie, C, *et al.*, 2012. First and repeated records of the tropical-temperate crab *Asthenognathus atlanticus* Monod, 1932 (Decapoda: Brachyura) in the eastern part of the Bay of Seine (eastern English Channel, France). *Cah. biol. Mar.* Vol. 53: 525-532

Kerckhof F, Haelters J, Gollasch S, 2007. Alien species in the marine and brackish ecosystem: the situation in Belgian waters. *Aquatic Invasions*. Vol. 2: 243–257

*Kraberg, A, Baumann, M and Dürselen, 2010. *Coastal phytoplankton: photo guide for northern European seas.* Verlag.

Le Hir,P, Bassoullet, E, Erard, E, Blanchard, M, Homan, D, Jegou, AM and Iriec, 1986. *Golfe Normano-Breton: etude régionale intégrée.* Ifremer: 6 vols.

Leppäkoski E, Olenin S, Gollasch S (eds), *Invasive Aquatic Species of Europe: Distributions, Impacts and Management.* Dordrecht, Boston and London, Kluwer Academic Publications

Lorenz J, Rauhut H, Schweitzer F, Helbing D, 2011. How social influence can undermine the wisdom of crowd effect. *Proceedings of the National Academy of Sciences.* Vol. 108: 9020–9025.

Lützen J, (1999. *Styela clava* Herdman (Urochordata, Ascidiacea), a successful immigrant to North West Europe: ecology, propagation and chronology of spread. *Helgoländer Meeresuntershungen.* Vol. 52: 383–391

Maggs CA, Guiry MD, (1987. An Atlantic population of *Pikea californica* (Dumontiaceae, Rhodophyta). *Journal of Phycology.* Vol. 23: 170–176

*Marine Biology Section, 2014. *The seashore life of Jersey.* Société Jersiaise.

*Martin, J, 2010. *Les invertébrés marins du golfe de Gascogne à la Manche orientale.* Éditions Quae.

MEA, 2005. *Millenium ecosystem assessment ecosystems and human well-being: biodiversity synthesis.* World Resources Institute.

Minchin D, Eno C, 2002. Exotics of coastal and inland waters of Ireland and Britain. In: Leppäkoski E, Gollasch S, Olenin S (eds), *Invasive Aquatic Species of Europe: Distribution, Impact and Management.* Kluwer, pp 267–275

Minchin D, Floerl O, Savini D, Occhipinti Ambrogi A, 2006. Small craft and the spread of exotic species. In: Davenport J, Davenport JL (eds), *The Ecology of Transportation: Managing Mobility for the Environment.* Springer.

Minchin D, Gollasch S, 2003. Fouling and ships' hulls: how changing circumstances and spawning events may result in the spread of exotic species. *Biofouling.* Vol. 19: 111–122

Minchin, D, Cook, EJ and Clark, PF, 2013. Alien species in British brackish and marine waters. *Aquatic Invasions.* Vol. 8(1): 3–19.

Mineur F, Johnson MP, Maggs CA, Stegenga H, 2007. Hull fouling on commercial ships as a vector of macroalgal introduction. *Marine Biology.* Vol. 151: 1299–1307

Nishikawa T, Bishop JDD, Sommerfeldt AD, 2000. Occurrence of the alien ascidian *Perophora japonica* at Plymouth. *Journal of the Marine Biological Association of the United Kingdom.* Vol. 80: 955–956

Orton JH, Winkworth R, (1928. The occurrence of the American oyster pest *Urosalpinx cinerea* (Say) on English oyster beds. *Nature.* Vol. 122: 241

Palmer DW, 2004. Growth of the razor clam *Ensis directus*, an alien species in the Wash on the east coast of England. *Journal of the Marine Biological Association of the United Kingdom.* Vol. 84: 1075–1076

Parrott D, Roy S, Baker R et al., 2009. *Horizon Scanning for New Invasive Non-native Species in England.* Natural England.

Ponder WF, 1988. *Potamopyrgus antipodarum* - a molluscan coloniser of Europe and Australia. *Journal of Molluscan Studies.* Vol. 54: 271–285

*Porter, J, 2012. *Bryozoans and hydroids of Britain and Ireland.* Marine Conservation Society.

Provan J, Booth D, Todd NP, Beatty GE, Maggs CA, 2008. Tracking biological invasions in space and time: elucidating the invasion history of the green alga *Codium fragile* using old DNA. *Diversity and Distributions*. Vol. 14: 343–354

Randall JM, Morse LE, Benton N, Hiebert R, Lu S, Killeffer T, 2008. The invasive species assessment protocol: a tool for creating regional and national lists of invasive nonnative plants that negatively impact biodiversity. *Invasive Plant Science and Management*. Vol. 1: 36–49.

Rilov G, Crooks J (eds), 2008. *Biological invasions in marine ecosystems: ecological, management and geographic perspectives*. Springer

Roy HE, Lawson-Handley LJ, 2012. Networking: a community approach to invaders and their parasites. *Functional Ecology*. Vol. 26: 1238–1248.

Roy HE, De Clercq P, Lawson Handley L-J, Poland RL, Sloggett JJ, Wajnberg E, 2011a. Alien arthropod predators and parasitoids: an ecological approach. *BioControl*. Vol. 56: 375–382.

Roy HE, Roy DB, Roques A, 2011b. Inventory of terrestrial alien arthropod predators and parasites established in Europe. *BioControl*. Vol. 56: 103–130.

Roy HE, Bacon J, Beckmann B *et al.*, 2012. *Non-native species in Great Britain: establishment, detection and reporting to inform effective decision making*. NERC.

Roy, HE, Peyton, J, Aldridge, DC *et al.*, 2014a. Horizon scanning for invasive alien species with the potential to threaten biodiversity in Great Britain. *Global Change Biology*. Vol. 20: 3859-3871.

Roy HE, Preston CD, Harrower CA *et al.*, 2014b. GB Non-native Species Information Portal: documenting the arrival of non-native species in Britain. *Biological Invasions*.

Ryland JS, de Blauwe H, Lord R, Mackie JA, 2009. Recent discoveries of alien *Watersipora* (Bryozoa) in western Europe, with redescriptions of species. *Zootaxa*. Vol. 43: 43–59

Ryland JS, Bishop JDD, de Blauwe H, El Nagar A, Minchin D, Wood CA, Yunnie ALE, 2011. Alien species of *Bugula* (Bryozoa) along the coasts of Atlantic Europe. *Aquatic Invasions*. Vol. 6: 17–31.

Sambrook K, Griffith, K and Jenkins, S, 2014. *Review of monitoring of marine non-native species in Great Britain and evaluation of gaps in data dissemination*. Natural Resources Wales. Report no. 20.

Shine C, Kettunen M, Genovesi P *et al.*, 2010. *Assessment to support continued development of the EU Strategy to combat invasive alien species*. Final Report for the European Commission. Institute for European Environmental Policy.

Smith P, Perrett J, Garwood P, Moore G, 1999. Two additions to the UK marine fauna: *Desdemona ornata* Banse 1957 (Polychaeta, Sabellidae) and *Grandidierella japonica* Stephensen 1938 (Amphipoda, Gammaridae). *Porcupine Newsletter*. Vol. 2: 8–11

Sousa R, Gutierrez JL, Aldridge DC, 2009. Non-indigenous invasive bivalves as ecosystem engineers. *Biological Invasions*. Vol. 11: 2367–2385.

Southward AJ, Hiscock, K, Kerckhof F, Moyse J, and Elifmov AS, 2004. Habitat and distribution of the warm-water barnacle *Solidobalanus fallax* (Crustacea: Cirripedia). *Journal of the Marine Biological Assossciation of the UK*. Vol. 84: 1169-1177

Stebbing P, Johnson P, Delahunty A, Clark PF, McCollin T, Hale C, Clark S, 2012. Reports of American lobsters, *Homarus americanus* (H. Milne Edwards, 1837), in British waters. *BioInvasions Records*. Vol. 1: 17–23

Stebbing, P, Murray, J, Whomersley, P and Tidbury, H, 2014. *Monitoring and surveillance for non-indigenous species in UK marine waters*. Cefas Report C5955(2)

Stubbings HG, 1950. Earlier records of *Elminius modestus* Darwin in British waters. *Nature*. Vol. 166: 277–278

Sutherland WJ, Woodroof HJ, 2009. The need for environmental horizon scanning. *Trends in Ecology and Evolution*. Vol. 24, 523–527.

Sutherland WJ, Fleishman E, Mascia MB, Pretty J, Rudd MA, 2011. Methods for collaboratively identifying research priorities and emerging issues in science and policy. *Methods in Ecology and Evolution*. Vol. 2: 238–247.

Thorp CH, Knight-Jones P, Knight-Jones EW, 1986. New records of tubeworms established in British harbours. *Journal of the Marine Biological Association of the United Kingdom*. Vol. 66: 881–888

Thorp CH, Pyne S, West SA, 1987. *Hydroides ezoensis* Okuda, a fouling serpulid new to British coastal waters. *Journal of Natural History*. Vol. 21: 863–877

Tidbury, H, Taylor, N, Copp, G, Garancho, E and Stebbing, P, 2014. *Introduction of marine non-indigenous species into Great Britain and Ireland: hotspots of introduction and the merit of risk based monitoring*. Cefas Report C5955(1)

Tidbury, H, Bishop, J, Stebbing, P, Yunnie, A, Wood, C and Sivyer, D, 2015. *The development and testing of methodologies for the use of an offshore buoy network for the early detection of bio-fouling non-indigenous marine species in the UK*. Cefas Report C5955(3)

Toupoint N, Godet L, Olivier F, Retière C and Fournier J, 2006. Does Manila Clam cultivation affect the *Lanice conchilega* bioherms of the Chausey Archipelago? *Benthic Ecology Meeting 2006, 35th Annual Meeting, Québec (Canada)*.

Trowbridge CD, 1998. Ecology of the green macroalga *Codium fragile* (Suringar) Hariot 1889: invasive and non-invasive subspecies. *Oceanography and Marine Biology Annual Review*. Vol. 36: 1–64

Verbrugge LNH, van der Velde G, Hendriks, AJ, Verreycken H and Leuven RSEW, 2012. Risk classifications of aquatic non-native species: Application of contemporary European assessment protocols in different biogeographical settings. *Aquatic Invasions*. Vol 7: 49-58.

Vilà M, Basnou C, Pysek P *et al*., 2010. How well do we understand the impacts of alien species on ecosystem services? A pan-European, cross-taxa assessment. *Frontiers in Ecology and the Environment*. Vol. 8: 135–144.

Willis KJ, Cook EJ, Lozano-Fernandez M, Takeuchi I, 2004. First record of the alien caprellid amphipod, *Caprella mutica*, for the UK. *Journal of the Marine Biological Association of the United Kingdom*. Vol. 84: 1027–1028

Withers RG, Farnham WF, Lewey S, Jephson NA, Haythorn RM, Gray PWG, 1975. The epibiota of *Sargassum muticum* in British waters. *Marine Biology*. Vol. 31: 79–81

Wolff, WJ, 2005. Non-indigenous marine and estuarine species in the Netherlands. *Zool. Med. Leiden*. Vol. 79 (1): 1-116.

Worsfold TM, Ashelby CW, 2008. Additional UK records of the non-native prawn *Palemon macrodactylus*. *Marine Biodiversity Records*. Vol. 1: e48

# Acknowledgements

We thank the following organisations for their assistance with the production of this guide: Alderney Wildllife Trust; Department of the Environment (States of Jersey); Guernsey Biological Records Centre; Ifremer; Jersey Biodiversity Centre; Jersey Seasearch; National Trust for Jersey; Seasearch (UK); Société Jersiaise; Société Guernesiaise.

We also thank the following individuals for their help with this and other marine biology projects: Lin Baldock; Louise Bennett-Jones; Jonathan Billot; Francis Binney; Samantha Blampied; Charlotte Boulton; Paul Chambers; Nina Cornish; Fiona Crouch; Sabina Danzer; Charles David; Chantelle de Gruchy; Denis de Gruchy; Andy Farmer; Anne Haden; Derek Hairon; Ed and Annie Hibbs; Charlotte Hooper; Courtney Huisman; Gareth Jeffreys; Nicholas Jouault; Phillip Langlois; Richard Lord; Dean Pitman; Roger Long; Richard Lord; Anya Martins; Kevin and Beverly McIlwee; Neil Molyneux; Greg Morel; John Pinel; Alex Plaster; Jon Rault; Jon Shrives; Jillian Smith; Mike Smith; Andrew Syvret; Bob and Jill Tompkins; Trudie Trox-Hairon; Kirk Truscott; Geoff Walker; Marion Walton; Chris Wood; Tim Wright; David Yettram.

# Image Credits

# Index to Species' Names